T0318841

Logistics Engineering and Health

Logistics Engineering and Health

Edited by

Hayfa Zgaya
Slim Hammadi

ELSEVIER

First published 2016 in Great Britain and the United States by ISTE Press Ltd and Elsevier Ltd

ISTE Press Ltd
27-37 St George's Road
London SW19 4EU
UK

www.iste.co.uk

Elsevier Ltd
The Boulevard, Langford Lane
Kidlington, Oxford, OX5 1GB
UK

www.elsevier.com

Notices

Knowledge and best practice in this field are constantly changing. As new research and experience broaden our understanding, changes in research methods, professional practices, or medical treatment may become necessary.

Practitioners and researchers must always rely on their own experience and knowledge in evaluating and using any information, methods, compounds, or experiments described herein. In using such information or methods they should be mindful of their own safety and the safety of others, including parties for whom they have a professional responsibility.

To the fullest extent of the law, neither the Publisher nor the authors, contributors, or editors, assume any liability for any injury and/or damage to persons or property as a matter of products liability, negligence or otherwise, or from any use or operation of any methods, products, instructions, or ideas contained in the material herein.

For information on all our publications visit our website at http://store.elsevier.com/

British Library Cataloguing-in-Publication Data
A CIP record for this book is available from the British Library
Library of Congress Cataloging in Publication Data
A catalog record for this book is available from the Library of Congress
ISBN 978-1-78548-044-7

Printed and bound in the UK and US

Contents

Hayfa ZGAYA and Slim HAMMADI

Hayfa ZGAYA and Slim HAMMADI

**Chapter 2. Case Studies and Contributions
to the Resolution of Logistics
System-related Problems**.   55

Hayfa ZGAYA and Slim HAMMADI

Chapter 3. Health Logistics: Toward Collaborative Approaches and Tools

Chapter 4. Collaborative Workflow for Patient Pathway Modeling at Pediatric Emergency Services

**Chapter 5. Agent-based Architecture for Task
Scheduling and Dynamic Orchestration Support** 139

Sarah Ben OTHMAN, Inès AJMI and Alain QUILLIOT

Preface

Recently, the academic and industrial world has understood that, in order to be successful, it is no longer enough to manage an organization relying exclusively on local performance objectives, in a compartmentalized way, but it is rather preferable to do it transversally and considering all possible options. Hence, the term "logistics", refers to all the activities that aim to select a set of resources at the lowest cost, for the purpose of conforming to customers' requirements. Nowadays, health-related logistics is accompanying demographic, socioeconomic and regulatory changes, particularly due to the appearance of new organization and evaluation modes. In this way, there is increasing awareness about the importance of the management of healthcare systems and of the control of different hospital flows. Indeed, actors from the hospital and healthcare sector must master the problems associated with process flows (i.e. patients, information, products, equipment) and with the internal restructuring manifested by the pooling of resources, in particular, that of technical platforms. However, healthcare professionals are neither prepared nor trained to solve such problems. It appears that they are lacking methodologies and decision support tools adapted to the demands that come with these future operation models. In this context, the current research and development works (R&D) aim to conceive and integrate innovative optimization and modeling methods that can provide decision support for the management of logistics systems. Such systems are often dynamic, spread over large-scale networks and presented as autonomous entities in interaction. The solutions that we recommend in this book provide answers to problems related to the management of logistics flows and the study of systems incorporating these flows. Owing to their increasing complexity, as well as their quality and speed decision requirements, it is necessary to respond to these challenges through an organizational, economic, technological and informational *innovative optimization* process, which will allow us to generate and perpetuate the necessary synergy to respond to the problems.

The works that we introduce in this book result from the fruitful collaboration between the Ecole Centrale de Lille (OSL team, OPTIMA group, CRISTAL CNRS UMR 9189)[1] and the University of Lille 2 (EA2694 Public Health Laboratory)[2].

Hayfa ZGAYA
Slim HAMMADI
June 2016

---

1 http://www.cristal.univ-lille.fr.
2 http://ea2694.univ-lille2.fr.

List of Acronyms

AFNOR:	French organization for standardization
AI:	Artificial Intelligence
AOA:	Agent-Oriented Approach
APS:	Advanced Planning and Scheduling
ASLOG:	French Logistics Association
BP:	Business Process
BPMN:	Business Process Modeling Notation
CAPM:	Computer-Assisted Production Management
CISIT:	International Campus for Transport Security and Intermodality
CMCG:	Consumption of Medical Care and Goods
DSS:	Decision Support System
DV:	Decision Variable
EA:	Evolutionary Algorithm
EC:	Equality Constraints
EDD:	Exchange of Digitized Data
ERP:	Enterprise Resource Planning
FL:	Flow Logistics
HES:	Hospital Emergency Services

HL:	Hospital Logistics
HSC:	Hierarchical Supply Chain
IC:	Inequality Constraints
ICP:	Industrial and Commercial Plan
IEA:	Integration and Evaluation Agent
ILP:	Integer Linear Programming
LN:	Logistics Network
LP:	Logistics Process
LS:	Logistics System
MAS:	Multi-Agent System
MPE:	Methods for Performance Evaluation
MSA:	Medical Staff Agent
NICT:	New Information and Communication Technologies
OF:	Objective Function
PES:	Pediatric Emergency Services
PIA:	Problem Identification Agent
PM:	Production Management
PMIS:	Program of Medical Information Systems
POA:	Process-Oriented Approach
RA:	Router Agent
RA:	Resource Agent
RDSO:	Regional Diagrams of Sanitary Organizations
RON:	Reception and Orientation Nurse
SA:	Scheduling Agent
SC:	Supply Chain
SCM:	Supply Chain Management
TA:	Tracing Agent

Introduction

There is increasing awareness about the importance of the management of healthcare production systems and the control of different hospital flows. This control aims to improve the quality of healthcare and is conditioned by health-related cost imperatives, risk management and quality. In recent years, this has been translated into different strategic and operational actions in the field of healthcare networks.

Numerous reports and studies describe the current state of hospitals and the healthcare system, which is undergoing a moral, demographic and financial crisis [COU 03a, MOL 05], a summary of which can be found in the work of Marcon *et al.* [MAR 08]. This situation is the cumulative result of new constraints and the strong rigidity of existing structures. In this way, hospital systems and emergency units are finding it more and more difficult to complete their missions.

Nevertheless, we observe that the French hospital and healthcare system is undergoing a global mutation. The healthcare culture, in particular public and private hospitals, is confronted with new concepts. Even though these initiatives may have a beneficial effect, the analysis of the necessary evolution of the healthcare system shows that a determining factor of this evolution consists of optimizing organizations, particularly their information systems. Indeed, many of the currently observed dysfunctions in hospitals and healthcare units result from an ill-adapted organization under the constraints and the evolution of their missions, as well as from the poor management of the flow of patients.

Introduction written by Hayfa ZGAYA and Slim HAMMADI.

The logisticians of these systems are confronted with problems of increasing complexity, such as: How is it possible to improve, secure and optimize logistics flows? How should the flow synchronization of these naturally dispersed systems, namely healthcare networks, global supply chains of multimodal transport, multi-site production networks and multi-zone crisis management, be improved? Which NICTs[1] should be adopted and how should they be harmoniously implanted taking into account the specificities of these logistics systems?

Thus, the actors from the hospital and healthcare system must solve those problems associated with process flows (i.e. patients, information, products, equipment) and with the internal restructuring that is manifested by the pooling of resources, mainly that of technical platforms. However, healthcare professionals are neither prepared nor trained to deal with such problems. It appears that they are lacking methodologies and decision support tools adapted to the demands that come with these future operation models.

This book presents research resulting from a fruitful collaboration between many CNRS research laboratories, health establishments and industrialists. This research contributes to the study and the development of logistics systems, in particular health-oriented logistics systems, in order to manage and optimize physical, informational and financial flows. In this book, we approach the problems surrounding the modeling, optimization and implementation of tools that help evaluate production, transport and crisis management systems, with a particular focus on health logistics systems. Nowadays, in the field of health logistics, the study and evaluation of overcrowding in hospitals (particularly in the emergency services) is becoming a major step toward finding efficient solutions that could improve patient care. These solutions represent the logistics engine of the healthcare establishment, which could gain in efficiency from the point of view of care, human resources and material management, medical activity pricing system and risk anticipation. A health logistics system must interact in a dispersed, uncertain and dynamic environment, enabling us to simulate, in a true-to-life way, the spread of logistics flows, the activities of the medical staff as well as their behavior and movements within the health centers.

---

1 New Information and Communication Technologies.

For example, the research works linked to the management of overcrowding in healthcare establishments have been supported and financed by the National Research Agency (NRA) in the context of the project NRA HOST[2] (2012–2015). This project HOST (Hospital: Optimization, Simulation and Tension Avoidance) proposes a solution for the implementation of a Decision Support System, making it possible to avoid the overcrowding that could take place at a healthcare establishment. More particularly, these research works deal with the question of how to better manage overcrowding at the *Services des Urgences Pédiatriques*[3] (SUP), which is the research domain of the HOST project.

The present book is organized into five chapters, as follows:

Chapter 1 presents logistics engineering by detailing the approaches of modeling, optimization and decision support to manage logistics flows.

Chapter 2 introduces the study of real cases of transport, management crisis and warehouse management logistics systems, specifying (for each field of application) the context, the problems studied and some possible solutions.

Chapter 3 is devoted to the study of hospital systems and emergency services. Here, we describe the healthcare sector: the context (international, national and regional), the hospital information systems as well as the hospital emergency systems. This helps us to describe the position of the hospital emergency services in the French healthcare system. Later, we will describe the functioning of pediatric emergency services.

Chapter 4 underlines the operational aspect of the hospital system, thanks to an innovative modeling approach, which allows the patient journey to be represented under the form of a collaborative workflow.

Chapter 5 introduces mathematical and algorithmic models of scheduling and dynamic orchestration of the collaborative workflow presented in the previous chapter by a multi-agent system. Many scenarios issued from real databases are tested and evaluated in order to anticipate and manage overcrowding of the pediatric emergency service.

---

2 http://www.agence-nationale-recherche.fr/?Projet=ANR-11-TECS-0010.
3 Pediatric Emergency Services (PES).

Logistics Engineering

1.1. Introduction

Today, in order to be successful, it is no longer enough to manage an organization relying exclusively on local performance objectives and in a compartmentalized way [LAU 04]. Rather, organizations have to be managed in a transversal way, considering all possible options. Hence, the term "logistics" [SOH 11] which has mainly been used in military language to refer to the art of combining all means of transport, supplies and housing facilities so as to conveniently manage the physical and information flows throughout a chain, called a logistics chain (LC), which starts at the provider's supplier and finishes at the client's customer. This is translated in terms of reliability and transfer speed of physical and information flows in order to attain the aims of the organization. In other words, logistics is interested in the planning and efficient follow-up of physical and information flows in order to satisfy a need. *It consists of all the activities that aim to select a set of resources at the lowest cost, for the purpose of conforming to customers' requirements.*

Logistics comes down to the following five questions:

– *What?* Which products to deliver?

– *How many?* In what quantities?

– *Where?* In which places?

– *When?* At which moments?

– *How?* With what means?

Chapter written by Hayfa ZGAYA and Slim HAMMADI.

The need for a logistics administration emerges from economic and social evolution. Today, this evolution needs innovative modeling, optimization and implementation techniques. The problem with the usual representation models of Logistics Systems (LS) derives from their inability to gather the structural and functional dimensions on the same diagram, which are by essence, transversal and complex [JOU 05]. These usual models exclusively juxtapose structural or functional bricks, but never both at the same time. For example, "actors" are never represented together with "activities". Today, in order to be competitive, an organization must have a precise and comprehensive view of not only its internal distribution but also the complete logistics environment in which it exists (suppliers, customers, competitors, etc.). All the elements must therein be present in order to take into consideration all of the parameters for local optimization of the structure and global optimization of the LS.

1.2. Logistics: origins and evolution

The word "logistics" has two origins: mathematical and military [LIE 07].

1.2.1. *Mathematical origin*

The origin of the word logistics is Greek: *logistikos*, which corresponds to mathematical reasoning. This term was first used by the Greek philosopher Plato (428–448 BC) and was the origin of the Latin word *logisticus*. It was first used in the French language in 1590 as an adjective to describe a logical reasoning.

At the start of the 20th Century, *Bertrand Russell*, British mathematician and logistician (1872–1970), highlighted the close link between logistics and mathematical logistics. This represented the beginning of the theory of algorithms. In this sense, logistics refers to the art of organizing a calculation in stages to attain an aim. This discipline is called algorithmic logics.

1.2.2. *Military origin*

The word "logistics" equally finds its source in the battlefields. Its meaning derives from the rank of an officer in charge of the dwellings of the troops during combat. Since ancient antiquity, logistics has played an important role in military activity. However, it was in the 20th Century, during the First and Second World Wars, that there have been important developments concerning the reflection in the practice of the subject, hence the development of a recognized science, called military logistics. The aim of this science is to regulate human resources, food and material flows in order to ensure, among others, the support in foodstuffs and the provision of equipment and transport means for the armed forces. The term "logistics" indicates, by an abuse of language, the science of the plan and execution of the transfer of armed forces and their maintenance.

1.2.3. *Evolution*

Logistics is an old word that has been used for centuries [MAT 10]. However, despite the fact that the term has evolved with time, it always refers to activities which aim to select the adequate resources (human, material, informational, etc.) to provide a service at the lowest cost.

Considered for a long time as the "logistics of modern times", logistics is today a general concept concerning all the operations that determine the movement of products. Many methodologies and organization methods have derived from this concept, the best known being "Supply Chain Management" (SCM) which marked the beginning of the 21st Century with its massive use by the professional and academic worlds (section 1.3.6). Basically, it deals with optimizing the LC in an organization. According to the *Association Française pour la Logistique*[1] (ASLOG), the aim of SCM is to coordinate the management of a network in terms of costs, deadlines and quality from the supplier's providers up to the distribution of these same goods to the final consumers.

In the 1940s, logistics mainly dealt with the functions related to the physical flows of distribution. However, owing to industrial development, the concept has evolved throughout the years, now including the evolution of internal and external factors to the firms.

---

1 French Association for Logistics, www.aslog.org.

Three great periods have marked the history of logistics (Figure 1.1) regarding its incorporation in Information Systems (IS): IS (section 1.6) of different logistics components (products, suppliers, customers, business orders, etc.). Thus, IS are classified into three types: separated (before 1975), integrated (after 1975) or cooperative (from the 1990s onwards). The presence of a logistics manager has become essential since the emergence of integrated IS:

– Separated logistics (service logistics): separation of all the logistics components.

– Integrated logistics (function logistics): integration of all the logistics components into a system. All the actors of this system communicate and possess a global view of the impact of logistics over the entire organization.

– Cooperative logistics (process logistics): the cooperation between the different logistics actors becomes important because of the insufficiency to integrate all the components under the same system. Cooperation creates real partnerships between suppliers, customers, competitors, etc., and participates in the emergence of communicative logistics tools such as Enterprise Resource Planning (ERP). The latter (section 1.6.3.2) deals simultaneously with the organization's internal and external data in order to better manage resources. Communication is done through the Exchange of Digitized Data.

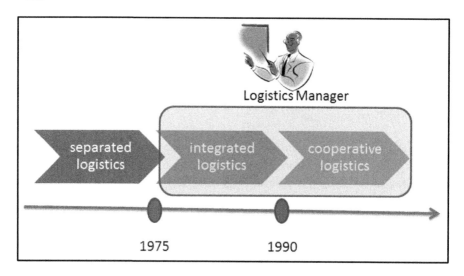

Figure 1.1. *Evolution of logistics*

1.3. Logistics Network: definitions, characteristics and complexity

1.3.1. *Logistics*

As mentioned previously, the concept of logistics has evolved through time. Today, many definitions exist according to different points of view. However, the most general definition is that of the ASLOG[2], which defines logistics as follows:

> Logistics comprises all the activities which aim to produce at the lowest cost, an amount of products to deliver, in order to meet a demand.

Logistics then has two complementary aims:

1) to make the necessary means available in order to ensure optimal flow conveyance (human, material and informational);

2) to efficiently secure Logistics Processes (LPs) (section 1.3.2) by managing resources and optimizing costs, in order to conclude logistics operations while sticking to deadlines.

1.3.2. *Logistics Process*

Process? Business Process (BP)? Logistic Process (LP)?

A *process* is an ordered sequence of operations, tasks, actions or activities responding to a certain diagram and leading to a result.

A *BP*, equally called operational process, involves business rules. These rules are the sources for decision (to perform the activities or not) at the heart of the process. BP activities are performed by actors (human or automatic) with the help of adapted means. The performance of each BP activity contributes to the achievement of an expected result. The business character of the process is equally expressed by the nature of the result, which must be effective, and should make sense to a final beneficiary. Nowadays, we talk of Workflow or Business Process Management (BPM), which corresponds to the electronic management of BPs (section 1.7.1.1).

---

2 French Association for Logistics.

A LP is a BP instance which aims to produce a result at the right moment, at the lowest cost and with the best possible quality service. It constitutes an operation and an event sequence which resorts to a set of means (human, material and informational resources) and controls the corresponding flows in order to achieve an objective at the lowest cost. These means can be human (staff), material (installations and equipment) or methodological (techniques and methods) and must be efficiently coordinated in order to attain the best possible service level. The LP crosses traditional organizational boundaries between the firms and their functions.

The X50-600 regulation of the *Association Française de la Normalisation*[3] (AFNOR) defines three deployment phases for a LP (Figure 1.2):

1) Planning: identifying the needs, conceiving the LS and developing it.

2) Execution: for the implementation of the conceived system and its development during the planning phase. This implementation must produce the desired result at the lowest cost while adhering to the deadlines.

3) Mastery: for flow control in order to identify the possible adjustments to be made at the previous phases, so that the system may continue to produce what it must, at the lowest cost while adhering to the deadlines.

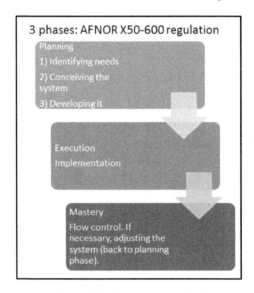

Figure 1.2. *The phases of a LP*

---

3 French organization for standardization.

1.3.3. *Logistics System*

A LS is associated with the assemblage of three categories of elements: 1) interactive actors, 2) resources used by these actors and 3) methods and tools employed to attain the aims of the system, enabling it to increase its performances:

1) actor: any unit playing a role in the LS: consumer, supplier, etc.;

2) employed resource: human, material or informational;

3) adopted method and tools: any employed means used for attaining the system's aims.

The aim of a LS is to provide services at the lowest cost, with the best possible quality, while adhering to the set deadlines. This production is possible thanks to human and material resources and grounding the procedure on the adopted methods and tools in order to warrant production efficiency.

LS can be spread over a distributed network. In this case, we would call it a Logistics Network (LN) or Supply Chain (SC) (section 1.3.5); otherwise, it would be a matter of internal logistics (section 1.4.1).

1.3.4. *Logistics Flow*

A Logistics Flow (LF) corresponds to any matter able to circulate throughout the components of a LS. Flow circulation can be made bottom-up to top-down or the other way round. We distinguish three categories of LF:

1) Physical flows: raw materials and finished products in the manufacturing process. The human flows can be considered as physical flows.

Flow direction:

i) Usual logistics: bottom-up→top-down

ii) Reversed logistics (retrologistics): top-down→bottom-up

2) Information flows: the exchanged information between the actors of a LS.

Flow direction: both directions are possible

i) bottom-up→top-down (e.g. acknowledgement of receipt of an order)

ii) top-down→bottom-up (e.g. quote)

3) Financial flows associated with physical flows.

Flow direction:

i) Usual logistics: bottom-up → top-down

ii) Refunds: top-down → bottom-up

In addition, these flows can be push or pull (Figure 1.3).

Push flows

Push flows refer to the top-down shipment of resources, from their reception at upstream zones (bottom-up), that is to say, from the supplier to the final customer. In this case, supplies are "pushed" the furthest toward the consumer. In military contexts, push flows correspond to the delivery of a resource considered enough for the field operator to provide.

Pull flows

Pull flows refer to a direct answer to a supply request, that is to say, to send the available resources when downstream (top-down) zones make the request.

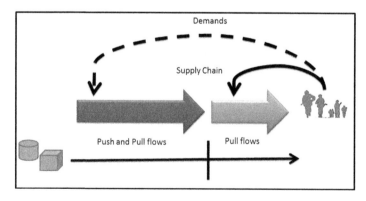

Figure 1.3. *Flow types*

1.3.5. *SC and LN*

"Men build too many walls and not enough bridges", Isaac Newton[4].

The objective of a LS is to have a transversal view over all interconnected logistics sites, by efficiently coordinating the actions of the actors involved. This makes it possible to reduce the dysfunctions resulting from a fragmented view of the circulation process of products, information and competencies. This approach can be made by crossing traditional organizational boundaries, be it between the organizations themselves or between their functions.

A LS can be represented in many ways, depending on the characteristics of the organization as well as the connections between its different actors and entities. Its best known form is the LC, deriving from the English term "Supply Chain" (SC). A SC is a set of hierarchically interconnected entities (sites, organizations, actors, etc.) These entities are called chain nodes and enable the circulation of push and pull flows. A LN is considered the most general form of SC, with arbitrary connections between the nodes (i.e. connections are not necessarily hierarchical). In general, we frequently refer to SCs rather than LNs, even though these are very widespread nowadays.

SCs involve all the service suppliers and customers who take part in the generation of a flow throughout the LP (section 1.3.5.2), from acquiring raw materials to delivering finished goods to the consumer.

One node in a SC can be a system representing a complete SC. Thus, one SC can be considered as a compilation of other interconnected SCs. For example, the healthcare SC (section 1.3.5.3) includes a blood donation SC, a drug SC and a hospital SC.

"SC" refers to a relatively recent concept, even if the military has used it for a long time. The most widespread definition has been provided by the Supply Chain Council[5], which defines SC as: "the successive production and

---

4 English philosopher, mathematician, physicist, alchemist, astronomer and theologian (1642–1727).
5 Non-profit association (founded in the Unites States in 1966) whose methodologies and analysis tools enable firms to significantly improve their SC management-related processes.

distribution stages of a product, from the provider's suppliers to the client's customers". A more general definition has been suggested by Poirier as: "A SC is the system which enables firms to bring their products and services to their customers".

From this global SC definition, we may deduce an even more general definition which acknowledges the fact that logistics is a concept involving many fields of action:

A SC corresponds to the system thanks to which the firms bring their products and services close to the final consumer.

Other definitions also exist in the specialized literature. Christopher [CHR 11] defines a SC as the "network of firms that are involved, bottom-up and top-down, in different processes and activities which create value in the form of products and services provided to the final consumer". In other terms, a SC is made up of many bottom-up (raw material and components) and top-down (distribution) enterprises, as well as the final customer.

Lummus [LUM 98] defines a SC as the "network of entities through which the material flow goes. These entities include suppliers, transport systems, assembly sites, distribution centers, retailers and clients".

1.3.5.1. *The case of a hierarchical SC*

A hierarchical SC (HSC) of a given level (L) is a particular LN composed of (n) node layers (representing the actors of the LN), in which the flow can only circulate between a pair of nodes at adjacent levels (Figure 1.3):

– n: the number of levels in the chain;

– $L=\{L_1, L_2,...,L_n\}$: chain's levels;

– n_i: the number of nodes of level i with $1\leq i\leq n$;

– $n_{i,j}$: the jth nodes of level i with $1\leq i\leq n$ and $1\leq j\leq n_i$;

– $O_i=\{O_{i,1}, O_{i,2},..., O_{i,ni}\}$: all the nodes of level i with $1\leq i\leq n$;

– the first level: the providers' suppliers;

– the last level: the clients' customers.

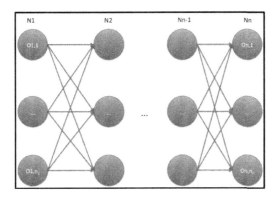

Figure 1.4. *General representation of a HSC*

For a HSC, any node chain level (i) can be linked to any other node of level (i−1) if 1<i≤n or (i+1) if 1≤i<n. When this hierarchy is not followed, that is to say, when any node can be tied to any other node in the system, we have the definition of a LN. The Crisis Management Supply Chain, introduced in our research works (section 2.4), is generally a HSC that enables resource management in case of a crisis.

1.3.5.2. *Example of a classical SC*

A classical SC is a HSC. Figure 1.4 shows a classical level 5 HSC (n = 5) with:

– L = $\{L_1, L_2, L_3, L_4, L_5\}$;

– number of nodes at each level: $n_1 = 2$, $n_2 = 2$, $n_3 = 3$, $n_4 = 3$ and $n_5 = 4$;

– the first level: raw material suppliers;

– the last level: final consumers.

1.3.5.3. *Healthcare SC examples*

The healthcare SC is a complex LS (Figure 1.6), which involves a variety of SCs, such as the drugs SC (Figure 1.7) and the blood donation SC (Figure 1.8). In this case, we have many types of products and services: blood bags, drugs, treatment, medical equipment, etc., and many consumer levels: healthcare centers, drugstores, etc., until we reach the final consumer (patient), the center of the healthcare SC.

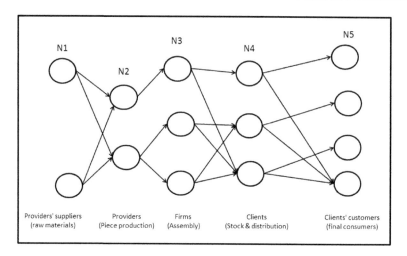

Figure 1.5. *Standard supply chain*

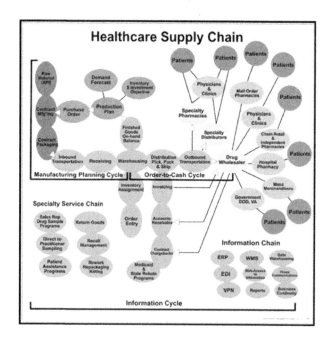

Figure 1.6. *Healthcare supply chain (source: Brentwood Management Group[6] – Healthcare Supply Chain Solutions)*

---

6 http://brentwoodllc.net.

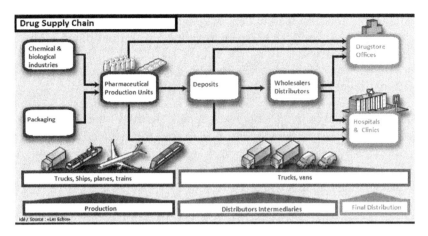

Figure 1.7. *Drug supply chain (source: [DU 10])*

A blood donation SC is a HSC. Figure 1.8 shows a level 4 (n = 4) blood donation HSC.

– L={L$_1$, L$_2$, L$_3$, L$_4$};

– the first level: donors;

– the last level: the patients benefiting from the blood bags.

This SC is stimulated by a four-stage process with two different circuits: the donor circuit and the blood bag circuit.

Stage 1: sampling

When a donor goes to a donation center, he completes a questionnaire about his lifestyle and medical history. Then, he sees a doctor who decides whether this person is a suitable blood donor. If yes, the sampling is performed by a specially trained nurse. We sample a blood bag that has to be prepared (stage 2), as well as sampling tubes that will be taken to a technical platform for testing (stage 3).

Stage 2: preparation

During the filtering phase, each sample bag is taken to a preparation platform in order to extract white blood cells. Once filtered, the bag has to be centrifuged in order to separate the components (red blood cells, plasma and platelets), because a patient is never transfused with total blood, only with the needed components.

Stage 3: biological characterization of the donation

This stage makes it possible to decide whether the blood can be transfused to a patient. On a technical platform, the sampled blood that has been previously stocked in special tubes goes through virological, serological and immunological tests. If the results show anomalies, the bag is discarded and the donor is warned.

Stage 4: distribution

Once prepared and described, the blood products are distributed to hospitals and clinics that have requested them, bearing in mind that blood products have different lifespans: 5 days for platelets, 42 days for red blood cells and many months for plasma, which can be frozen.

The donation is always performed according to compatibility criteria (immune-hematology) in order to prevent the aggregation of the antibodies of the receiver with the antigens of the donor. Many controls must be performed in order to choose the blood that best corresponds to the receiver's needs.

These stages can be modeled in the Workflow (Chapter 4) in order to better manage the coordination between the different actors of the blood donation SC.

1.3.6. *Supply Chain Management*

1.3.6.1. *Definitions*

The concept of SCM was suggested by Olivier and Webber in 1982. This concept, complementary to that of SC, refers to the approaches, processes and functions that are essential to the efficiency of an organization, reducing its costs and increasing its flexibility. The aim is to optimize SC performances thanks to an optimized coordination of its components.

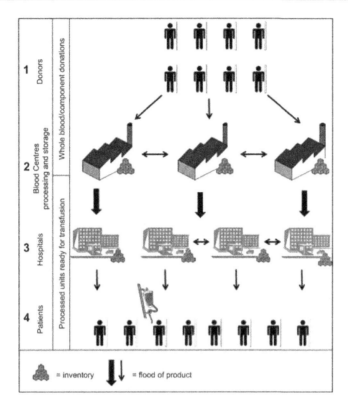

Figure 1.8. *Blood donation supply chain (source: [STA 11])*

Despite the fact that many authors insist on the difficulty of defining the concept of SCM [MEN 11], we will collect a variety of definitions in this book, mainly extracted from Croom *et al.* [CRO 00] and Wolf [WOL 08].

Vakharia [VAK 02] defines SCM as "the art and science of creating and improving the synergic relations between the partners of a same SC with the common aim of delivering, in time, the pertinent products and services, to the right customer, with the best quality".

Simchi-Levi [SIM 99] describes this concept as the efficient use of many approaches for the best possible integration of the SC actors. These take part in the different stages of the LP. The aim is to deliver the final product, in the right quantity, at the right moment, at the lowest cost and with the best possible service quality for the final consumer. Simchi-Levi [KAM 03], in his book, equally suggests the following definition: "SCM is a strategy that aims simultaneously at the reduction of global expenses, providing a more competitive position to the different parts of the SC, as well as optimizing final customer satisfaction for a greater adaptability of production and distribution systems".

For Rota-Franz [ROT 01], "SCM" means that we aim to integrate the variety of internal and external means to respond to the customer's demand. The aim is to optimize all LPs simultaneously, not sequentially.

All these definitions allow us to understand the meaning of SCM by its aim: to improve competitiveness by minimizing costs, ensuring good customer service and efficiently allocating activities to actors in the production, distribution, transport and information sectors, checking that these actors do not develop local antagonistic behaviors that could affect the global performance of the system.

All these definitions describe SCM as the management of physical and informational flows aiming to optimize orders, production and delivery LPs. We thus stick to the following definition:

> SCM involves optimizing SC management. It consists of efficiently managing the total scope of the firm's LPs, from the providers' suppliers to the clients' customers.

Therefore, the management of LS requires developing models and optimization methods that improve efficient decision-making. Indeed, LS generally constitute particularly difficult socio-technical organizations and present modeling and optimizing complex problems. The LS we mainly refer to are those related to production, transport, health and crisis management. Such systems are often dynamic, diffuse and extended over large-scale networks and generally take the form of interactive autonomous entities. The processes resulting from these systems are complex, because of their considerable size (important number of variables), the nature of their dynamic relations and the multiplicity of constraints to which they are subjected (productivity constraints and security for human and the environment).

1.3.6.2. *SCM complexity*

Many issues prove the broad complexity of SCM:

– Aims are often in conflict for the actors involved (e.g. customer response, flexibility, lot sizes, inventories, costs).

– SCs are dynamic systems, for example, because of the evolution of markets, actors and the relations between them.

– Systems vary in time (e.g. production planning, strategic pricing, availability, supply costs).

– SCs are complex networks of enterprises, industries and organizations that may have conflicting aims.

– Balance between offer and demand is very difficult to obtain.

– Conception and management problems are new, badly understood by the actors and standard solutions are unavailable.

– The volatility of demand and inventories increases top-down. This phenomenon, which must be constantly prevented and reduced is called the "bullwhip effect" [TOR 06].

Nowadays, informational tools are necessary for counterbalancing the complexity of SCM. Over many time scales (strategic, tactical and operational), SCM must be based on a collection of (informational) tools destined for piloting the SC of an organization and optimizing the working of its components. These tools are integrated into Informational Systems (IS), specifically called Informational Systems for Logistics (ISL) (section 1.6). Before moving on, we will introduce the possible classifications of logistics as well as of attainable logistics performances, thanks to the implementation of such IS.

1.4. Logistics typology

Logistics can be classified from different perspectives.

1.4.1. *According to space*

1) Internal logistics: in this case, the aim of logistics is to place an amount of goods (or a certain number of services) in an internal environment

(the same node of a SC or a LN) to satisfy a demand in the most efficient way (according to chosen criteria). For example, in the healthcare context, this could refer to Pediatric Emergency Department or the internal organization of an operation room.

2) SC logistics: in this case, the aim of logistics is to place an amount of goods (or a certain number of services) to satisfy a demand in the most efficient way (according to chosen criteria), but this time in a distributed manner along the different nodes of a SC or a LN. It is often a question of finding the right balance of flow transfer in order to ensure quality logistics.

1.4.2. *According to the field*

A non-exhaustive list of different logistics fields is provided below:

– industrial/production logistics;

– humanitarian/crisis management logistics;

– health logistics;

– agri-food logistics;

– transport logistics.

1.4.3. *According to the function*

A non-exhaustive list of different logistics functions is given below:

– Distribution logistics: performed in the top-down direction of a SC, gathers exit flows and concerns final consumers.

– Supply logistics: performed in the bottom-up direction of a SC, gathers entry flows and concerns suppliers.

– Warehouse logistics: administrates *picking*[7] and the management of availabilities.

– Stock logistics: stock management describes when and how much it is necessary to supply by limiting stock and shortage costs. Many techniques exist. In production logistics, the calculation is made through the Material

---

7 Picking goods from their stock furniture in order to prepare an order.

Requirement Planning (MRP) method, which makes it possible to calculate net needs. This method is integrated into SC management tools such as ERP (section 1.6.3.2).

– Reverse logistics: equally called return or retro-logistics, for the management of refunds.

– Service logistics: to suggest, with the right timing, the best possible service to a user according to his preferences.

1.4.4. According to the decision-making levels of an organization

Here, logistics is characterized in terms of the decision-making horizon, that is, the necessary perspective in order to make decisions: long, middle, short and very short term.

1) Piloting logistics: this concerns the strategic level. It particularly deals with long-term decisions such as the choice of providers and subcontractors, the number and location of warehouses, etc. This decisional level is defined in section 1.5.3.1.

2) Support logistics: this concerns the tactical level. It deals with mid-term decisions, such as activity planning in relation to available resources. This decisional level is defined in section 1.5.3.2.

3) Operational logistics: this concerns short-term decisions, as in stock management and machine programing. These are production and scheduling decisions that are limited in relation to time and space. This decisional level is defined in section 1.5.3.3.

4) Execution logistics: this concerns very short-term decisions, even real-time decisions, for example, launching Manufacturing Orders (MOs) during the production process. This decisional level is defined in section 1.5.3.4.

In the following sections, we will develop these different decision-making levels of logistics further, highlighting the quality of results produced by decisions made in the context of high-performance LS. Nowadays, we talk of the convergence of quality and logistics, which illustrates the tight connection between these two.

1.5. Quality/logistics convergence

Convergence is a term used in different fields. According to *Larousse* dictionary, it means tending toward the same point, the same objective or the same result (convergence of efforts). In geometrics, we speak of the convergence of two lines. In physics, convergence refers to the property of luminous rays tending toward the same point. In mathematics, it is the property of a series whose term sum tends toward a finished limit when the number of terms increases indefinitely. In algorithms, a program *converges* if all the values it engenders tend toward a desired limit in the considered research space, satisfying optimality conditions [ELK 12].

In order to improve the performances of an organization and to reach fixed objectives, it is necessary that all the actors of the firm efficiently coordinate their actions, hence the expression "quality/logistics convergence". According to C. Hohmann[8] [HOH 06], the convergence of quality and logistics increases the performance of the global system and offers an interesting potential to economies.

These actions result from short-, mid- and long-term decisions, which make it possible to perpetually improve the performances of organizations. The good decisions to be made are often difficult to identify because of the complexity of LS. In the following sections, we describe the difficulties as well as the right decisions to be made in order to improve the performances of such systems.

1.5.1. *Logistics performances*

The implementation of a system measuring performances means the need for controlling the LS in order to achieve fixed objectives. Logistics performance is measured with performance indicators. These indicators are sometimes difficult to identify and quantify. Beamon [BEA 99] classifies these performance indicators into two categories: qualitative performance measures (customer satisfaction, flexibility, integration of physical flow and information, financial risk management, etc.) and quantitative (late deliveries, time of customer response, etc.). As quoted above, according to C. Hohmann, the global performance of a LS is measured by the result

---

8 Manager at *Agamus Consult*, expert in "Lean Manufacturing", conducting research into logistics performance through continual improvement and waste elimination.

of the trio quality/costs/deadlines. These three are the most widely used performance indicators.

Resulting from these measures, re-engineering decisions must be made. The aim is to act over the system via Decision Variables (DVs) in order to tend toward the fixed objectives and to increase the performances of the system (Figure 1.9).

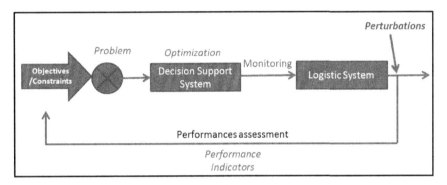

Figure 1.9. *Control system for the performances of SCs*

1.5.2. *Difficulties associated with the performance improvement of a LS*

In his book, Siarry [SIA 14] raises a variety of difficulties to be overcome for optimizing the performances of a LS in order to make it more competitive. These difficulties are basically classified into two categories:

1) Modeling difficulties

A LS is difficult to model because there are many interacting entities that have to be represented (e.g. actors and activities). These entities are often subject to management rules, which are difficult to define. The model construction has to be carried out with an appropriate finesse degree considering the decision-making level and with regard to the desired objectives. In section 1.7.1, we explain the concept of logistics modeling as well as possible solutions for efficiently representing a LS.

2) Optimization difficulties

LS are often big particularly because of the significant number of actors, products and services. This implies an algorithmic complexity of the programs that seek the best possible solution for optimizing criteria. These are often contradictory, for example cost minimization (transport, stock, production, etc.) and the maximization of the customer service rate (service quality). The concept of logistics optimization as well as the possible solutions to optimize a LS are introduced in section 1.7.2.

1.5.3. Decision-making

Many types of decisions must be made in a LS according to perspectives: long, middle, short or very short term. The longer the horizon, the more the uncertainties. A LS must adapt and remain performing in the face of uncertainty, hence the need to study the reactivity of the system in case of eventual hazards in order to manage risk: epidemics, natural disaster, broken machines, etc.

Decision-making at the interior of the system is done at the level of its different processes and entities (e.g. supply, production, distribution and sales). These decisions can be made horizontally or vertically (see Figure 1.9) [LEM 08]:

– Horizontal decision-making corresponds to the synchronization of different entities of a LS in the same horizon. For example, a plan for a site (such as production) cannot be carried out because of the constraints of another site of the same LS (such as supply). It is then important to ensure that the plans of different sites are coherent.

– Vertical decision-making corresponds to the organization of decisions in time for the same site.

As shown in Figure 1.10, vertical decision-making is generally divided into three levels: strategic, tactical and operational [BOT 05, GAN 99, SHA 99, VIN 04]. Each level interacts with others and corresponds to a decision horizon. There exists a fourth level: executive, which corresponds to very short-term decision-making, even in real time (Table 1.1). These four decision levels correspond to Global Fulfillment [COH 01], which is an integrated approach of enterprise processes that make it possible to better respond to customers' demands. An approach that gathers all decision-making levels is necessary to efficiently conceive and pilot a LN [KOU 06].

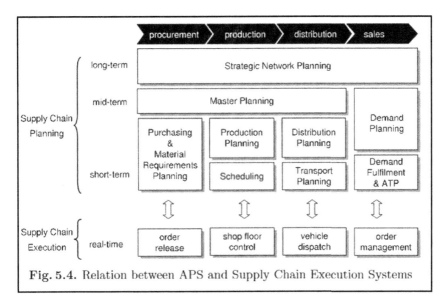

Fig. 5.4. Relation between APS and Supply Chain Execution Systems

Figure 1.10. *Strategic, tactical and operational levels in a SC (source: [ROH 00])*

Decisional level	Horizon	Time frame
Strategic	Long-term	From 6 months to many years
Tactical	Mid-term	From some weeks to some months
Operational	Short-term	From some hours to some days
Executive	Very short-term	From real time to some hours

Table 1.1. *SC decision-making levels*

1.5.3.1. *Strategic level: "the long term"*

Strategic plans concern a relatively long-time perspective, from several months to some years (generally 2–5 years). This level, also called "strategic management" by Croom *et al.* [CRO 00] or "strategic planning" by Thomas and Griffin [THO 96], gathers all strategic decisions, particularly the conception of LNs, for instance, the problem of site location, well known under the name of Facility Location Problem. These decisions involve not only long-term directives and action lines, for example, the search for new industrial partners, the choice of suppliers and subcontractors, but also others, such as implanting new intervention zones (in the case of military logistics or their delocalization), appointing a new supply zone to a distribution center (warehouse), developing a new product, configuring SCs, coordinating their function modes as well as defining the financial objectives to be attained.

1.5.3.2. Tactical level: the "mid-term"

This level corresponds to the Industrial and Commercial Plan for a family of products and services and concerns a middle-term horizon, varying from some weeks to some months. This level deals with mid-term decisions that must be executed in order to deploy the strategy, which has been decided by the firm. Decisions mainly concern problems associated with the management of available resources and are based on the management of flows, such as activity planning depending on available resources during a time frame or the lot sizing problem, calculated with Wilson's formula in practice [BRU 14]. At this level, tactical decisions generate a detailed production plan or schedule (flow shop problem, job shop problem, etc.) to be applied at an operational level, for example, at a workshop or a logistics zone.

1.5.3.3. Operational level: "the short term"

This level, equally called "operational planning" according to Thomas and Griffin [THO 96], corresponds to the Production Directing Plan, more accurately than the IPC. It concerns a relatively short-time horizon, from some hours to some days (decisions are to be made in a day or along the week). These decisions have a limited spatio-temporal impact.

1.5.3.4. Executive level: "the very short term and real time"

This level corresponds to very short-term, even real-time, decision-making. In the very short term (only a few hours maximum), we may calculate Net Needs[9] and ensure their control and piloting in real time. For instance, during the production process, this concerns issuing a MO.

1.5.4. Discussion

Decision-making in Information LS (ILS) must permanently improve logistics performances. In order to make the best possible decisions, the logician must receive the proper information at the right time. He also has to

---

9 Net Need Calculation (NNC).

feed the necessary data into the system that will enable future local or global decisions. This is one of the most important roles of an ILS.

1.6. Infologistics: information systems for logistics

According to Marouseau [MAR 05] and the model of Venkatraman [VEN 95], the concept of IS emerges from the convergence of technological and competitive pressures. New logistics means appear, which aim at improving the performances of LS. At present, we talk about Infologistics, a term combining information and logistics. It concerns all the tools and technological solutions that improve the optimization and information control of logistics flows along a LS. These means are integrated into the ISs of the firms.

1.6.1. *Introduction*

An IS comprises all the resources for stocking, processing, managing and distributing information in a firm. Resources can be of many types: data, procedures, personal, material and logistics.

An ISL is an IS that contributes to efficiently piloting SCs. Nowadays, ISLs are spread over four layers (Figure 1.10), each of them corresponding to a decision-making level (section 1.5.3), as quoted above.

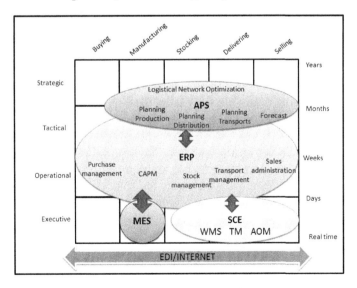

Figure 1.11. *Information systems for logistics*

1.6.2. *Evolution of ISL*

Production Management (PM) comprises all the activities taking part in the conception, resource planning (material, financial and human), scheduling, registering and controlling of production in an enterprise. The aim of this discipline is to optimize organizational processes by continuously improving the flows going from suppliers to customers. All these activities must be performed (implicitly or explicitly) following procedures established by the company and take into consideration not only the quality of its products and services, but also the security of its employees and its environment.

In their book, Courtois *et al.* [COU 03b] have highlighted the evolution of PM toward logistics, which covers a wider range of activities dealing with the integrated management of flows: "Management production has progressively evolved towards an activity which aims at coordinating and circulating the flows of value creation over a larger perimeter, at the fastest speed, resorting to adjusted and optimized resources, and dynamically ensuring that demands meet customers' deadlines". IS have evolved due to their heterogeneity (very short/short/mid/long term) and their expansion in time and space.

Currently, PM manipulates a large amount of data. Consequently, and to efficiently manage their operational processes, firms employ computerized tools, integrated into their IS. These include not only Computer-Assisted Production Management (CAPM), but also ERP, as well as supervising tools. These computerized tools have evolved within PM-related activities to cover those that contribute to the integrated management of flows (Figure 1.11). Indeed, while it has long concentrated around CAPM and scheduling or production control software, PM has witnessed the emergence of new systems (ERP, SCM, APS, MES, etc.) which have surpassed the administrative sphere to offer a panel of transversal computer tools for different decision-making levels. This evolution corresponds not only to an evolution in the basic functions integrated into the CAPM software, but also to the integration of related functions, which have considerably modified the extent of industrial management. This explains its evolution toward global logistics.

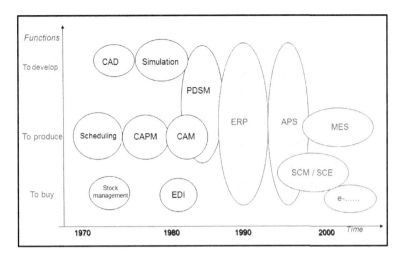

Figure 1.12. *Evolution of information systems for logistics*

1.6.3. *Software resources of ISL*

An ISL is composed of software layers destined to different decision-making levels (Figure 1.12). There are many types of software layers: Advanced Planning and Scheduling (APS), ERP, Manufacturing Execution System and Supply Chain Execution (SCE). Some examples of the use of these tools according to their logistics function and decision perspective are presented in Table 1.2.

1.6.3.1. *APS: "Advanced Planning and Scheduling"*

Since the mid-1990s, APS has played an innovative decision support role in comparison with other existing software on the market. It consists of a special type of software package that makes it possible to optimize planning and synchronizing of physical flows in a SC, by taking into account a number of possible constraints: resources, capabilities, deadline costs, etc. An APS is based on data issuing from other layers of the ISL.

The aim of an APS is to satisfy a demand with available resources. To achieve this, it models constraints, expresses cost functions and researches the values of DVs by optimizing criteria [COU 03b]. Currently, optimization is done thanks to resolution engines based on linear programing or by constraints, but research continues to improve these resolution engines by integrating advanced approaches to optimization and modeling, such as metaheuristics and Multi-Agent Systems (MAS). (section 1.7.1.2).

Thus, an APS makes it possible to improve the reactivity of companies not only when being used internally (scheduling or need calculation) but also at the global level of the SC. For example, an APS offers the better choices, the best possible appointments, the best stock levels, the most appropriate subcontractors, according to chosen criteria.

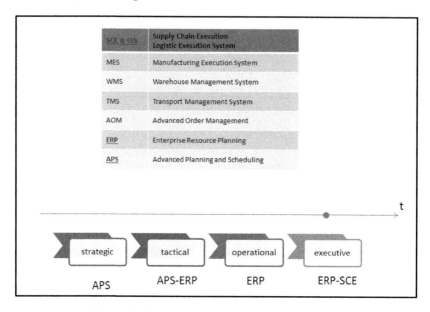

Figure 1.13. *Tools according to decisional levels*

1.6.3.2. *ERP: "Enterprise Resource Planning"*

ERP is a professional tool that integrates all the functions of a company to efficiently manage all its processes. ERP contains, among others, a PM module based on the MRP method. It is a system that appeared during the 1970s, which calculates Net Needs in resources according to sales forecasts and product classification. The aim is to satisfy internal and external demands (as expressed by customers), by synchronizing the amount of raw material and semi-finished products (physical flow synchronization). MRP has evolved into Manufacturing Resource Planning (MRP2), which makes it possible to make decisions depending on time requirements (horizon). As MRP2 is a mono-site technique (local logistics), Thomas and Lamouri [THO 00] consider that SCM is the equivalent of MRP2 in multi-sites.

Even though MRP is considered as an "ancestor" of ERP, it is still one of its fundamental modules. In the 1960s, MRP was used to transform commercial data from sales into production facts, whereas nowadays MRP2 integrates other functions, such as infinite planning, production control and cost calculation.

As it is a package application, ERP consists of different modules that cover all the needs of an industrial and commercial enterprise, by ensuring transparency, traceability and exchange optimization. On the other hand, ergonomics and the navigation system must be common to the functions of ERP. The majority of ERPs currently present a client/server or Internet/Intranet architecture [CHE 04].

Therefore, the main interest of an ERP is to replace the different applications of an IS by a unique, coherent and homogeneous system (e.g. impeding multiple data capture) and eliminate the long and expensive developments of homemade applications. Composed of many modules, an ERP practically supports all the operational processes of a firm, and allows maximum coverage: financial and accountancy management, management control and instrument panel, PM (MRP, Manufacturing plans), purchase management, supply policies, stock management, sales administration, invoicing, prospective marketing, delivery and distribution logistics, human resources and salaries, etc.

To summarize, ERP is a professional management software with the following properties:

– practically supports all the operational processes of an enterprise (maximum coverage);

– supports a unique and integrated view of operational information (maximum computerization degree);

– makes information transfer transparent and immediate throughout different functions (real-time ubiquity);

– is provided and kept by a unique editor/producer (maximum degree of technological computerization).

1.6.3.2.1. The market of ERPs

With 10 million users worldwide, a € 6,265 billion turnover in 2000 and more than 20,000 installations, Software Package Application Systems (SAP) is an ERP leader controlling 45% of the market share. Created in Germany in April 1972, this software package has evolved with the appearance of many versions: SAP R/2 in 1979; SAP R/3 in 1992; MySAP.com, e-business in 1999, etc. The success of this product can be explained thanks to the immediate value of every operation (consuming, movement, etc.) linked to its real-time online function.

The most widely used ERP training tool is PRELUDE, an educational software for initiation and deep learning of SC mechanisms.

1.6.3.3. SCE: "Supply Chain Execution"

SCE makes it possible to integrate all the information related to the SC's operational management activities. It is generally composed of three interfaced management applications: TMS, WMS and AOM.

– TMS: "Transport Management System", for optimizing the organization and cost of transport circuits.

– WMS: "Warehouse Management System", for the management and optimization of warehousing operations.

– AOM: "Advanced Order Management", for the management and administrative processing of orders and promotions.

Decisional level	Software	Functions				
		Buying	Manufacturing	Stocking	Transporting	Selling
Strategic	APS	Which suppliers?	Which factories? Which subcontractors?	Which distribution network?	Which transport means? Which transporters?	Which products and services? Which customers?
Tactical	APS and ERP	Purchase planning	Production planning	Distribution planning	Transport planning	Sale forecast
Operational	ERP	Purchase management	Production management	Stock management	Transport management	Sales administration
Executive	MES and SCE	Supply	Workshop control	Warehouse management	Circuit management	Order entry
			MES	SCE – WMS	SCE-TMS	SCE-AOM

Table 1.2. *IS for the decision-making levels of an enterprise (from* Stratégie logistique *(no. 63, nov. 2003))*

1.6.3.3.1. ERP/APS complementarity

An ERP does not take external constraints into consideration and plans its needs only as a function of demands or demand forecast, that is, considering an unlimited production capacity always capable of providing, without being able to react to hazards (breakdowns, delivery delays, etc.), hence the need for APS as a decision support tool. An APS intervenes during SC synchronization. It acts as a support for decisions and planning for the whole chain by using the information provided by an ERP. In fact, ERPs are transactional tools that ensure data capture and stock (e.g. order entry, product declaration), whereas APS software types are tools specialized in structuring decisions for demand planning, production and distribution [GEN 05].

APSs and ERPs are essential and complementary tools in an IS. Their cohabitation in an IS must be successful. APSs must be able to determine the best decisions by taking into account the most recent information, and also to ensure plan controls and make sure the situation is still on track. An APS formats the information entered on the ERP so that the manager can establish a plan. Production decisions, transport and supply information are then transmitted to the ERP in order forms so as to be launched [GEN 05].

In planning terms, APS scan replace ERP planning modules by making a global plan of the total SC in its materials, components and products. They can equally make an optimal planning of the elements judged critical (Figure 1.13). The plans for other articles will be executed in the ERP by the MRP constraints.

APSs give visibility to all the chain and offer network functionalities with ISs of customers and providers thanks to Intranet and Internet technological developments. Today, the SC manager can rely on a complete informational architecture, enabling him to make decisions based on upgraded data, thanks to ERP and with the support of ABS.

More and more, ISL base their operations on advanced computerized methods and tools that we shall present in the following section.

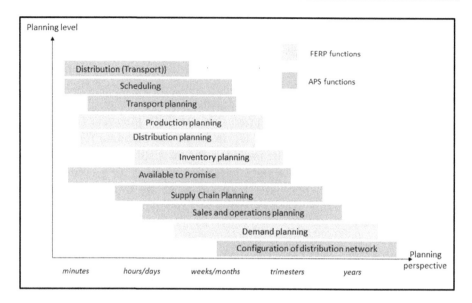

Figure 1.14. *ERP/APS complementarity. For a color version of the figure, see www.iste.co.uk/zgaya/logistics.zip*

1.7. Possible resolution methods in favor of logistics

In order to improve the performances of a LS, it is necessary that all its actors coordinate their actions in such a way to improve the total result, making the best possible decisions at every moment. It is one of the major roles of an IS, because it allows us to gather and manage more and more important masses of information thanks to the use of tools such as ERPs. In fact, while it is true that the tools which have been now developed are able to integrate the means – still basic – that enable decision support (opting in each case for the most appropriate expert vision), this is achieved at the cost of some simplifying hypothesis: horizon choice, degree of data liability, definition of evaluation criteria, etc.

Innovative means and decision support tools today integrate two principles: modeling and optimization.

Modeling refers to the means which represent an organization, that outlines its functioning and that reveals its available resources, as well as the constraints to which it is subject. It is then a way of representing the global architecture or a part of a LS as well as the interactions between and inside parties. Thanks to modeling, the organization can be tested, simulating normal functioning (permanent mode) or perturbed functioning (transitory mode). Modeling enables system simulation before it is implemented and it also makes it possible to observe and, if necessary, to correct the LS according to the chosen management parameters.

Finally, coupled with this modeling, optimization tools can equally be used. The aim of this complementarity is to clearly represent a system (via a model) before optimizing its functioning. Indeed, each actor of the LS must make decisions that will have an impact on the global performance of the system. These decisions must be effective, and therefore, it is necessary to use an optimization approach. Optimization represents the way of integrating behavior in the most efficient way, globally or locally in each part of the system. Simulation makes it possible to have an idea about the functioning of the system according to a given configuration.

These new tools, considered as innovative logistics tools, make it possible to model complex systems, by considering the behavior of actors, flow exchange, resource use, etc., and to optimize the functioning of the system by making a note of its constraints.

1.7.1. *Approaches to modeling and simulation*

Modeling allows us to represent a system in order to visualize a schematic reproduction of it. Figure 1.14 (taken from Abo-Hamad and Arisha [ABO 11]) shows the modeling procedure of a real LS based on its specifications. Modeling makes it possible to verify the functioning and robustness of a system before really building it. This verification is done through simulation, which corresponds to scientific experiences partially or totally performed thanks to computerized tools. Thus, simulation becomes a rational and essential step for scientific research. In the same way that the Latin expression "*in vitro*" describes how to make a living experience, simulation constitutes "*in silico*" experimentation, that is to say, research or tests performed through the use of computer technologies.

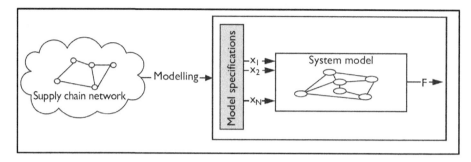

Figure 1.15. *Representation of a LS using the modeling method (from [ABO 11])*

There are many methods and tools available for the modeling and the simulation of a LS. The choice is made according to its properties as well as the target or aim of modeling.

Indeed, modeling methods are Methods for Performance Evaluation (MPE) that take into account the structural and functional complexity of a LS. A MPE must consider the system uncertainties, its constraints and its possible contradictory criteria, such as cost reduction and quality service increase. The result of a MPE is a model for performance evaluation (ModPE).

These methods and tools are used for modeling and analyzing a LS in the field of health [FAR 11].

An approach that can efficiently represent LSs is the Process-Oriented Approach (POA), which considers the different entities of a LS as interdependent processes. Two types of POA exist: the Agent-Oriented Approach (AOA) and Workflow [ZGA 09, BOR 05]. In comparison with Workflow, AOA is more oriented toward the rational and social components of a system.

1.7.1.1. *Workflow approach*

The best known POAs for the modeling of organizations are: Architecture of Integrated Information System (ARIS) [KEH 11, SHE 02], Supply Chain Operations Reference (SCOR) [HUA 05] and Analysis, Specification, Conception and Implementation (ASCI) [GOU 91]. However, the tendency is now oriented to process modeling that visually represents the activities of organizations in the form of "active flowcharts". This discipline uses the acronym BPM, for "Business Process Management". It consists of providing an organization with the means to pilot and master its BPs and not only the means to model them. BPM takes its name from the GRAI method [DOU 98], which mainly focuses on the decisional aspect of the process. A more recent approach, corresponding to a new type of modeling, has emerged. This approach is called Workflow.

At first created for describing, representing and implementing the BPs of enterprises (section 1.3.2) with the aim of efficiently organizing their activities [MAR 04], Workflow is often used for modeling and simulating scientific processes, which is why it is often called scientific Workflow.

This approach makes it possible to represent a system or to implement it using a graphical language inspired from Unified Modeling Language (UML) and based on four component groups: activities (to-do lists), events (important events that trigger, finish or link processes), transitions (allowing flow circulation) and conditional connections (for flow orientation). Business Process Modeling Notation (BPMN) is a Workflow graphical language which has been developed by BPMI[10] [CHI 12], merged with Object Management Group, which founded UML in 2005.

Workflow makes it possible to create a model representing the behavior of a system and to later implement it thanks to an integrated engine. During the implementation, Workflow allows human actors to interact, at the right moment, reading and writing with the system thanks to graphical interfaces. During the reading phase, in order to benefit from the right information at the right moment to carry out ongoing tasks, and during the writing phase, to feed the system with the necessary information at the right moment to oversee its smooth running.

The most recent research works in the field of scientific Workflow seek to improve and to facilitate the transformation of abstract models from their conceptual level (user's schematic representation according to study area and objectives) to their concrete level (computerized implementation of the model) [CER 13].

---

10 Business Process Management Initiative.

Software using BPMN language is a Workflow tool. These are often equipped with simulation modules in order to check that system processes are running correctly, for example, examining resource consumption and turnaround time. An interesting comparison of these tools was carried out by Liu *et al.* [LIU 11]. For example, apart from having a complete free download version, Bonitasoft[11] has a model implementation feature that can generate interfaces which invite human resources to interact with the system and to simulate model implementation thanks to a built-in simulation tool. It is equally possible to link Workflow with other external systems thanks to special connectors.

1.7.1.2. *Agent-Oriented Approach*

Agent-Oriented Approach (AOA) is a modeling methodology mainly used in the context of complex systems. It includes POA because an agent is considered as an "intelligent" process [ZGA 07a], in the form of an autonomous entity behaving rationally and with a "social life". In an agent society, an agent may develop a "strategy" corresponding to all the actions that characterize the behavior of the agent. An agent is then capable of interacting with other agents in a MAS, based on interaction protocols.

1.7.1.2.1. When should the AOA be used?

AOA is used in two scenarios, complex systems simulation and problem resolution:

– System simulation: the aim of the simulation is to observe and to understand its behavior as well as the behavior of its components in a stochastic environment.

– Problem resolution: the aim of the problem resolution is to implement a Decision Support System. When it is a question of splitting a complex problem into many sub-problems, we often resort to a top-down approach. Each sub-problem is then represented by many autonomous entities, even rational ones. The latter will solve the global problem in a cooperative way thanks to their interaction.

1.7.1.2.2. Agent: definition and characteristics

An agent is an autonomous (acting alone or with the help of other agents) and rational (with many intelligence levels) entity, representing a real or virtual component and displaying a variety of behaviors. An agent has a

---

11 http://www.bonitasoft.com.

lifecycle (Figure 1.15): first, it perceives its environment (via its sensors); then, it upgrades its internal condition as a result of this perception; after, it reasons, and finally it acts (thanks to its effectors). It can be created at any moment without disrupting the functioning of the global system, and can also be cloned, provided with mobility, can be in a waiting state, destroyed and can even commit suicide (destroy itself).

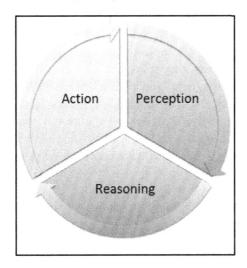

Figure 1.16. *Lifecycle of an agent*

The rationality of an agent represents its intelligence. This intelligence is associated with its granularity[12] level: reactive reasoning (stimulus → response), cognitive reasoning (reasoning based on mental attitudes such as belief, desire and intention) or hybrid reasoning (reactive and cognitive). As that of the human being, the reasoning capabilities of an artificial agent are not necessarily static. Agents can learn from their experiences and from their interactions with their environment or with other agents or objects belonging to another environment.

1.7.1.2.3. Multi-Agent System

An agent can interact with local agents in a centralized environment or with distant agents in a broader environment, constituting a MAS. An agent may have one of many roles in a system. To play a role, the agent adopts a

12 The level of detail considered in a model or decision-making process. The greater the granularity, the deeper the level of detail.

variety of behaviors associated with this role. Thus, a behavior may adopt different forms: parallel, sequential, composed, complex, etc.

How should a successful MAS be achieved?

The conception of agent-based systems constitutes a research field on its own. The common interest to all existing methods is to represent in an efficient way the different behaviors.

In order to achieve a successful MAS, we resort to an agentification procedure. It is a process that transforms an existing system in a set of autonomous interacting agents. The efficacy of the AOA mainly depends on this agentification process. The greatest difficulty in creating an efficient model is to find a compromise between the global problem and the local distribution of agents, on the one hand, and the number of agents and the volume of information to be treated (granularity level of agents), on the other hand.

Globally, there are two approaches toward multi-agent modeling:

1) Top-down approach: the global system is dissociated into many sub-systems, each of them being represented by an agent (Figure 1.16). In this case, there is explicit inter-agent coordination.

2) Bottom-up approach: the agents are identified and reunited forming a coalition which represents the global system (Figure 1.17). In this case, there is emerging inter-agent coordination.

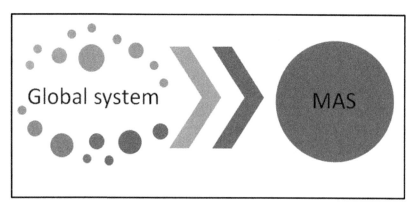

Figure 1.17. *"Top-down" multi-agent modeling approach*

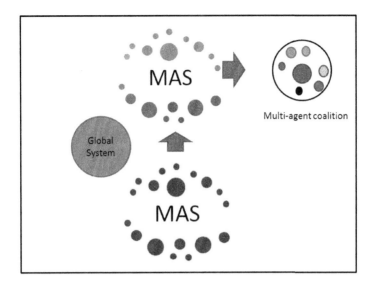

Figure 1.18. *"Bottom-up" multi-agent modeling approach*

Coalition concept

A coalition can be defined as a short-term organization based on specific and contextual engagements, which enables agents to benefit from their respective competencies.

Agent modeling methods

There are many agent-oriented modeling methods (e.g. O-Mase). For instance, the MAS team of the informatics research laboratory at the University of Pierre and Marie Curie (CNRS UMR 7606) is interested in the conception of MAS and, in particular, of cognitive MAS. This team focuses on the formal specifications of MAS models. In fact, formal specification of a MAS allows for the verification and, consequently, the validation of a system before its implementation. The team has created a formal specification method of cognitive agents based on the cognitive model of reference for beliefs, desires and intentions (BDI) [BRA 87]. This method, called CTL_AgentSpeak, is integrated into the programing language oriented to agents: AgentSpeak [GUE 09].

1.7.1.2.4. Emergence of agent-based systems

Once the MAS has been conceived and implemented, it evolves according to the internal behavior of the different system agents as well as their interaction protocols. In complex systems, we can place the agentification procedure bottom-up of the emergence phenomenon. Sure enough, emergence is defined as the unexpected production of new entities, results or structures resulting from the interaction of entities in the system according to a dynamic bi-directional repetitive process (*top-down* ←→ *bottom-up*). Figure 1.18 represents the emergence phenomenon according to Ueda [UED 01], who explains it as follows: *A global order of structure expressing new function formed through bi-directional dynamic process; where local interactions between elements reveal global behavior and the global behavior results in new constraints in the behavior of the elements.*

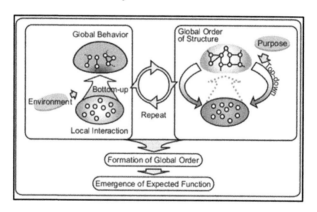

Figure 1.19. *The phenomenon of emergence in complex systems (from [UED 01])*

1.7.1.2.5. Interaction and convergence of a MAS

Interaction is a dynamic relationship established between many agents resulting from their combined and reciprocal actions [DON 06]. It can have different forms: cooperation, coordination, negotiation, collaboration, etc. Nevertheless, it is basically represented by the coordination of all the actions that allows us to achieve the system's objectives. In Weiss [WEI 99], coordination is defined with regard to shared resources because agents must coordinate their actions in the environment they share in order to avoid conflict over resources.

In this context, we say that a MAS converges when all the objectives of the agents are achieved; otherwise, it diverges (i.e. all the possible means to achieve convergence have failed). To attain these objectives, there are two types of coordination (Figure 1.19):

– when the agents behave as antagonists, they are competitive. Interaction then acquires the form of negotiation protocols;

– when the agents are not antagonists, they must implement resource-sharing strategies in order to avoid conflict. Interaction then takes the form of collaboration protocols representing a cooperation method that enables work distribution. Two types of complementary categories can be implemented:

- centralized planning: at the interior or the agent accounting for his behavior (the agent has to decide on the best way to interact with other agents),

- distributed planning: at the level of the MAS, who act according to collaboration protocols with other agents.

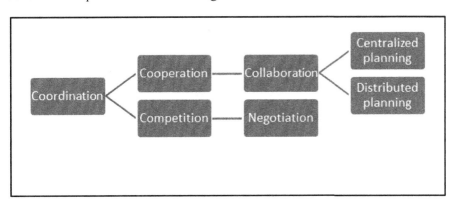

Figure 1.20. *Different inter-agent interaction forms*

The form of interaction is represented by a protocol. Many standard protocols exist in the literature, for example, the *Contract Net Protocol* [HSI 06]. A variant of this protocol, proposed by Zgaya and Hammadi [ZGA 07b] enables to accelerate the convergence of MAS with partial agreements.

1.7.1.3. *Agent-based simulation*

Owing to their ability to realistically reproduce complex environments, MAS have enticed researchers' curiosity, to study and analyze the growing phenomenon inside these environments, thanks to simulation.

Simulation of a multi-agent environment is based on the three main stages of an agent's lifecycle (section 1.7.1.2.1):

1) environment perception (the part of the environment that the agent can perceive);

2) upgrading of the agent's condition in response to his perception;

3) (after reasoning) selection and choice of the interaction that the agent will follow.

However, recent research in the field of agent-based simulation categorically depends on the characteristics of the simulated environment, as well as their entities' behavior. In [LAC 13], the authors put forth a first proposal for formalizing an agent-based generic simulation system, but this system demands a predefined reasoning plan for simulation.

CRISTAL's team of Multi-Agent and Behavior System (SMAC) specializes in the construction of open experimentation platforms for the simulation of agent-based environments.

One of the latest realizations of this team is the Ioda/Jedi[13] project, a package including a modeling method and a multi-agent simulation tool centered on the interactions that this method allows. In the context of this project, the agent simulation takes place in discrete time intervals, because every passage from one environment state to the other does not necessarily require the passage through a sequence of intermediary states.

In this manner, the project is composed of two main parts: the modeling methodology for simulation (called IODA: "Interaction-Oriented Design of Agent simulations") and the simulation display unit (called JEDI: "Java Environment for the Design of agent Interactions"), based on the concepts of the IODA methodology.

---

13 http://www.lifl.fr/SMAC/projects/ioda.

Another creation of this MAS team is the ATOM[14] project, "Artificial Open Market". This project, in collaboration with the IAE[15] of the University of Paris I, is a multi-agent simulation environment for financial markets.

1.7.2. *Optimization approaches*

1.7.2.1. *Introduction*

The optimization of the LS is essential in order to have the best possible performance at the lowest cost. This optimization process does not only involve the system in its totality, but must also consider optimization of each local component. In this context, we should consider three optimization levels [SIA 14]:

– functional: refers to coordination of different activities (IS, warehouse management, production logistics, product conception and lifespan, etc.);

– temporal: making decisions according to time frames (section 1.4.4). However, the more important the horizon, the more it is necessary for the LS to remain adaptable and highly proficient in the face of uncertainty;

– geographical: making decisions for mono-sites and multi-sites.

Despite the fact that optimization has ancient origins, it was actually developed at the turn of the 20th Century, with the appearance of the concept of stock management (together with the economic lot formula). We then looked for the means and techniques for decision-making in view of obtaining the best possible result, that is, we developed decision support tools and, in that way, Operational Research (OR) was born. Its development after the Great War provided organizations with the means to deal with their logistics problems through a quantitative approach. The aim was to succeed in the analysis and diagnose of complex situations in order to assess risks and make adequate choices. In this context, decision-makers have to face problems of increasing complexity, which derive from very diverse technical sectors, such as logistics, planning, service organization, etc. [JAC 12].

---

14 http://atom.univ-lille1.fr.
15 *Institut d'administration des entreprises de Paris.*

1.7.2.2. *Definitions and concepts*

1.7.2.2.1. Optimization methods

Optimization methods are techniques that enable us to solve optimization problems. For this, it is necessary to optimize the working of a system, minimizing or maximizing one of its many objectives or performance criteria.

1.7.2.2.2. Optimization problems

An optimization problem comes down to finding the optimal value – called *optimum* – corresponding to the *minimum* or *maximum* of an Objective Function (OF). This value depends on a set of DVs chosen to optimize a certain number of criteria (time, costs, etc.). The choice of these DVs must be done with respect to a set of equality constraints (EC) or inequality constraints (IC). All these constraints define the research space for the optimal solution. Formally [COL 02, DRE 03], an optimization problem is presented as follows:

Minimizing $f(\vec{x})$ (the OF to optimize) with:

– $\vec{x} \in \mathbb{R}^n$: comprises the (n) of DV

– $g(\vec{x}) \in \mathbb{R}^m$: gathers the (m) of IC

$$g(\vec{x}) \le 0$$

– $h(\vec{x}) \in \mathbb{R}^p$: gathers the (p) of EC

$$h(\vec{x}) = 0$$

1.7.2.3. *Typology of optimization problems*

Many types of optimization problems exist, depending on the features of the problem, the number and type of DV, the type of OF and the type of problem (formulated with or without constraints). This classification is summarized in Table 1.3 [ZGA 07a]. Thus, one problem may have different types. For example, human and material resource allocation problems in a medico-social center can be multi-variable, combinatory, nonlinear, with constraints (adhering to deadlines, not overconsuming resources) and multi-objective (optimizing time and workload). When DV are real continuous numbers, we refer to a continuous optimization problem; otherwise, the optimization problem is combinatory, operating over discrete variables.

Problem characteristics			Type of problem
Decision variable	Number	1	Mono-variable
		>1	Multi-variable
	Type	Real continuous number	Continuous
		Whole	Discrete
		Permutation of a given set of numbers	Combinatorial
Objective Function	Type	Linear function of DV	Linear
		Quadratic function of DV	Quadratic
		Nonlinear function of DV	Nonlinear
	Objective number	1	Mono-objective
		>1	Multi-objective
Problem formulation	type	With constraints	With constraints
		Without constraints	Without constraints

Table 1.3. *Classification of optimization problems [ZGA 07a]*

In the field of OR, the problems are strategic (how to find the best solution as a function of different constraints and objectives). More precisely, we refer to operational difficulties such as scheduling problems (to determine the scheduling order of a number of tasks according to constraints and objectives), stock management, (human and material) resource allocation for tasks, etc.

1.7.2.4. *Optimization method typology*

In order to solve an optimization problem, it is important to carefully model it, then to apply an ad hoc algorithm that corresponds to the chosen resolution method. Resolution methods for combinatorial optimization problems are globally classified into two categories: exact methods and approximate methods. The latter are used when exact methods are limited, for example, due to the size of research space or the impossibility to model the problem in a linear way.

1.7.2.4.1. Exact methods and approximate methods

Exact methods allows us to obtain and prove the optimality of the best solution thanks to an exhaustive research with an explicit enumeration of all possible solutions. These methods refresh techniques coming from integer linear programming (ILP), such as *branch-and-bound, branch-and-cut, Lagrangian Relaxation* and backtracking algorithms.

Nowadays, generic software for ILP is available (AMPL, CPLEX, LINDO, MPL, OMP, XPRESS, etc.). These applications make it possible to easily solve problems that can be written in an algebraic form, either in binary or in whole variables.

While these methods are practical for solving reasonable size problems, they are not recommended for solving difficult optimization problems, due to their time-consuming nature. Moreover, considering that the necessary calculation time to find a solution may rise exponentially depending on the size of the problem, exact methods generally encounter difficulties with a large volume of applications, even if these methods have brought real progress to the field.

For example, for the well-known OR Traveling Salesman Problem (for a salesman who has to visit a given number (n) of cities exactly once, following the shortest possible route and going back to the origin city), the research space increases in $(n-1)!$ Thus, for 50 cities, it will be necessary to evaluate 49! possibilities. This results in a combinatorial explosion.

Consequently, while it is true that exact method resolutions allow us to obtain one or more solutions with guaranteed optimality, in certain cases, we can be satisfied with a good quality solution (not the optimal), but in benefit of a reduced calculation time. For that, we use an approximate method.

In general, two reasons incite us to employ approximate methods: 1) the impossibility to solve the problem in a reasonable time due to the large size of data and 2) the impossibility to model the problem in a linear way.

There are two categories of approximate methods: heuristic and metaheuristic.

1.7.2.4.2. Heuristics and metaheuristics

Heuristics (from the Greek verb *heuriskein*, which means "to find") is an optimization method generally conceived for a specific problem, in order to produce an optimal (or near-optimal) solution [DRE 03]. In fact, heuristics does not generally provide an optimality guarantee, even if we have been able to prove convergence in some cases. These approaches are often based on a stochastic principle non-deterministically in order to overcome combinatorial explosion.

Metaheuristics refers to a more general method, similar to an algorithmic "toolbox", ready to use in different optimization problems, and requiring only a few modifications so that it can be adapted to particular problems[16]. In concrete terms, it can be defined as a set of concepts used for representing heuristic methods, which are applicable to a great variety of problems. Sometimes, it profits from experience accumulated during the search for an optimum, and it resorts to learning strategies in order to better guide the rest of the process in the quest for optimal solutions [OSM 96].

1.7.2.4.3. Metaheuristics

Metaheuristics have a major role as decision support tools for complex LS [LOU 01] composed of sites and organizations that have interconnected activities and diverse (and generally contradictory) objectives. It is like a great family of optimization methods seeking to solve very complex problems, such as combinatorial optimization problems with discrete variables and global optimization problems with continuous variables. These methods are characterized by a high level of abstraction, which enables them to adapt to a set of problems, ranging from simple local search algorithms to complex global search algorithms.

There are many advantages concerning the use of metaheuristics in decision support LS:

– modular nature:

  - simplicity and implementation facility;

  - robustness;

  - short development and maintenance time frames (this is an advantage over other techniques for industrial uses).

---

16 See metaheuristics.org.

– capacity for solving complex problems and managing uncertainty: these methods have already proved their efficiency in resolving difficult optimization problems:

- manipulation of large information volumes;

- complex problems are not necessarily simplified and must be treated as such in order to better represent reality.

– the possibility of managing uncertainty by studying many real scenarios of the complex problem instead of suggesting an exact but simplified global solution for a problem whose main data are the result of estimations.

There are many types of metaheuristics:

– local research method: for example, simulated annealing and Tabu search;

– constructive approach: to reduce the size of the problem at each stage, with the aim of progressively restricting the scope of possible solutions;

– evolutionary approaches:

- evolutionary algorithms: genetic algorithms, dispersed research, etc.,

- ant colony algorithms,

– hybrid approaches: the combination of evolutionary algorithms with local research algorithms can produce interesting results. This combination may actually increase efficiency in the resolution of complex problems.

As there exist no generic solution models, approaches for conceiving solutions to complex problems demand a great deal of analysis. For example, the solutions for genetic algorithms present themselves in the form of chromosomes. The efficiency of problem resolution will then depend on the efficiency of the chromosome model.

For instance, Zgaya [ZGA 07a] proposes a chromosome called flexible tasks assignment representation (FeTAR) (Figure 1.21). FeTAR has a matrix shape in which the columns represent service suppliers and the lines represent the services to provide. Each matrix cell represents allocation of services to the supplier. There are three possible values: "X" if the supplier does not propose the service, "*" if the supplier proposes the service but has not been chosen to provide it and "1" if the supplier has been chosen to provide the service. The purpose is to find the best possible supplier allocation, knowing that suppliers can propose the same services at different costs, qualities and response times (service processing time).

	S_5	S_{18}	S_1	S_{14}	S_{201}	S_{50}	S_8	S_9	S_{10}	S_3
T_{12}	*	*	*	1	*	*	*	*	*	*
T_3	*	*	*	*	X	*	*	*	*	1
T_5	1	*	*	*	X	*	*	*	*	*
T_6	X	X	X	X	X	1	X	X	X	X
T_{27}	X	1	*	X	X	X	*	*	X	*
T_{10}	X	X	*	*	X	X	*	*	1	X
T_{32}	*	*	*	1	X	X	X	X	*	X
T_{13}	*	*	*	*	X	X	*	*	*	1

Figure 1.21. *FeTAR chromosome for optimizing service allocation (from [ZGA 07a])*

1.7.3. *Modeling and optimization for the benefit of logistics*

In the context of a horizontal and vertical synchronization of flows and activities and in order to evaluate and improve the performances of a LS and its competitiveness, a global vision should be adopted (section 1.5.3). In this context, the system's complexity can be considered as a combination of many optimization-interfering problems.

Numerous works show the interest of combining optimization methods and modeling methods to solve complex problems. Modeling makes it possible to represent a system by considering its structural and functional complexity. At the core of the system (represented by a model), optimization problems (central or diffuse) are solved thanks to one or various optimization methods. Resulting solutions (values given to DV) are introduced into the model representing the system.

For this, there are two procedures, usually complementary:

1) To introduce the given solutions into the model to verify, or even prove, their efficiency without real implementation. In this case, the model is

called the simulation model and the different actors of the system are virtual [ABO 11].

2) To introduce the given solutions (supposed optimal) into the model, with real implementation. In this case, the model represents a real-life system, physically set, and the different actors in the system are real.

In order to have a robust and efficient system, it is necessary to model it (taking into account its structural and functional complexity). In this case, modeling represents the system's architecture. Otherwise, optimization algorithms – representing the system's behavior – are integrated into the model in order to optimize it. Once virtually implemented in the first place, the working of the virtual system is verified thanks to simulation. If the simulation result is satisfactory regarding expectations, the system can be really implemented. If not, it is necessary to calibrate the different modeling or optimizing parameters of the system before launching simulation, and so on.

Figure 1.22 shows the complementary character that exists between optimization and modeling for a virtual LS. In this case, we are referring to a simulation model. Indeed, the most widespread tool for the validation of decentralized systems (and particularly of complex LNs) is simulation. The most popular method of the simulation validation consists of building a system model, and then, comparing the results generated by this system to the real situation, and evaluating them following different performance key indicators.

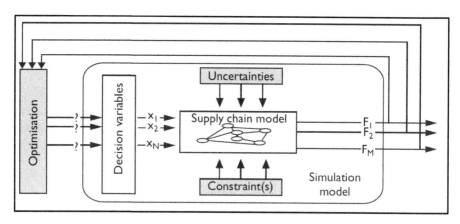

Figure 1.22. *Interaction between optimization and modeling (from [ABO 11])*

Current problems of LS

LS represents highly combinatorial optimization problems of exponential complexity. As a consequence, it becomes necessary to decompose these systems into a set of sub-systems of minimum complexity and maximum interconnection.

In a "multi-threading" context, these sub-systems are associated with more or less independent processes, executed in parallel and in an autonomous and interactive manner.

In this context, research approaches adopted to solve complex logistics problems by the OSL team of CRISTAL are based on the alliance between MAS and optimization.

In fact, an agent can have many roles in a LS. To play a particular role, the agent adopts a set of behaviors. For example, to solve an optimization problem, the agent plays an optimizing role and adopts one or more behaviors associated with this role. For example, to anticipate overcrowding at an emergency service, an agent can play an anticipating role and adopt one or more behaviors associated with this role.

There are two resolution approaches corresponding to agent modeling approaches, introduced in section 1.7.1.2:

– the global problem of optimization is dissociated into various sub-problems, given to each agent, created in real time (top-down approach);

– agents are identified and they gather forming a coalition to attain a common aim: the resolution of a global problem (bottom-up approach).

According to the type of interaction (cooperation, coordination, negotiation, etc.) between the different agents of the same system and the available resources, the system can converge toward a reference logistics solution thanks to a collaborative optimization process between agents. To put in practice this collaborative approach, the OLS team of CRISTAL proposes a generic model composed of three layers, in which the ground layer depends on the studied LS.

The three-layer model at the service of LS

The aim of the research presented herein is to identify a reference logistic situation through the use of mathematical and algorithmic models. In the case of environment perturbation associated with the real logistic situation, the three-layer architecture model suggested in Figure 1.23 has an objective to quickly attain the reference logistic situation, thanks to a collaborative optimization approach between the agents of the second layer. This architecture is called A3C-2SL for *Architecture à 3 Couches au Service des Systèmes Logistiques*[17].

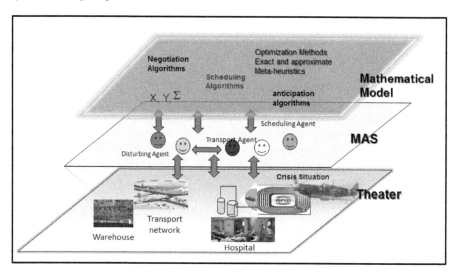

Figure 1.23. *A3C-2SL architecture for optimizing LS*

The different layers of A3C-2SL architecture are detailed as follows:

– the first layer composed of mathematical and algorithmic models (exact methods, metaheuristics, etc.) enabling to characterize the reference logistic situation;

– the second layer based on communicating agents that enables to identify the real ground situation. The reference logistic situation should be attained thanks to the collaborative optimization process between agents.

17 Three-Layer Architecture at the Service of LS.

– the third "layer" representing the real logistic situation. This varies according to the field: crisis management logistics, transport logistics, hospital logistics, warehouse logistics, etc.

This architecture, innovating and generic, is based on communicating agents who represent the different actors in the logistics chain and who are in direct relation with what happens on the ground. These agents continually observe the information on the ground layer by comparing the real situation with the reference logistic situation. According to this information, as well as diverse mathematical models (available at the first layer), these agents must adapt their roles and behaviors in order to better react to different ground disturbances, with the aim of attaining the reference logistic situation as quickly as possible.

1.8. Conclusion

In this chapter, we have introduced part of the basic elements of a LS, on the one hand, and the innovative approaches for their management and resolution, on the other hand. We have highlighted the interest of a three-layer resolution architecture in the service of LSs (A3C-2SL). This architecture represents the backbone of the research activities of the OSL team of CRISTAL laboratory (UMR CNRS 9189) and enforces a collaborative optimization approach essential for better management of the current LS: in health, transport, warehouse and crisis management.

Case Studies and Contributions to the Resolution of Logistics System-related Problems

2.1. Introduction

In this chapter, we present the contributions and logistics collaborative optimization research applied to the fields of transport, health, crisis management and warehouse management. For each of the mentioned logistics areas, we introduce a study of the context, the problem involved as well as different possible resolution approaches. All these fields have many points in common in terms of modeling and optimization approaches of tools.

2.2. Analogies between logistics systems

All the logistics systems in the fields of transport, health, crisis management and warehouse management share many points in common (Figure 2.1). In fact, all these systems need to be modeled and optimized. In the cases where they evolve in multi-criteria environments (generally with antagonistic or conflictive criteria), these logistics systems equally need decision support tools.

Chapter written by Hayfa ZGAYA and Slim HAMMADI.

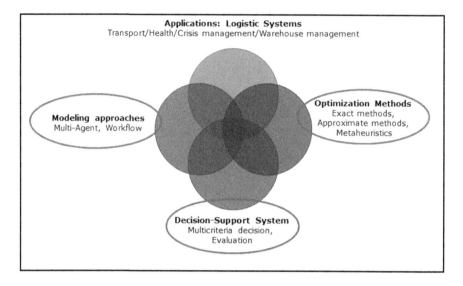

Figure 2.1. *Analogies between logistics systems*

In the following sections, we introduce study cases concerning logistics problems in different fields (ground layer variation: Figure 1.23), showing the similarities and the methodological adaptation to efficiently manage these systems.

2.3. Transport logistics

2.3.1. *Context*

The studies associated with logistics in the field of transport can be classified into two categories: *regulated transport* and *customer transport.*

Regulated transport. In urban transport networks, the exploitation of a line is divided into two complementary phases. The first phase deals with conceiving and elaborating the production program, which engenders a basic movement table. This phase corresponds to planning, which is done in anticipation of the exploitation period, and is necessary not only to schedule vehicles, assigning them to each course, but also to schedule the staff, assigning them to services. The second phase is regulatory and is done in real time. It consists of adapting the production program to the real exploitation conditions, aiming at satisfying different objectives, such as punctuality, regularity and connections. Besides, the production of collective transport operates in a completely random universe, basically because of the disturbances associated with rolling stock or circulation uncertainties (traffic jams). In this context, the regulator needs a Decision-Support System (DSS) in order to offer efficient regulatory strategies.

Customer transport. Studies carried out in this field aim to provide transport customers with all the necessary information for their journeys, without being obliged to connect with many databases to gather required information. The aim is not only to spare the traveler from looking for information which is difficult to find, but also to provide the traveler with the optimal service that best suits his expectations: the best rate, maximum comfort, minimum connections, etc. The purpose is to accompany the transport customer from his departure point to his arrival point, before, during and after his trip, by offering a rich and varied choice concerning not only transport and itineraries, but also the associated services that may orient him during his trip and also constitute entertainment.

2.3.2. Problems

The difficulty is basically associated with the growth of the volume of information on computer networks, which is increasingly extended and distributed and thus generates a commercial contest for information.

Research works in this field solve the problem by using models and informatics optimization of multimodal and co-modal transport systems. Multimodal transport corresponds to itineraries which combine many means of transport, such as train, bus, subway and tramway. Co-modal transport is an extension of multimodal transport, which does not aim at competition but rather complementarity (Figure 2.2) between public transport, private transport and free service vehicles (VLib, autoLib, etc.).

In this context, research works of the OSL team of CRISTAL aim not only to optimize the multimodal and co-modal itineraries [JER 12, KAM 07, WAN 15, ZID 06] but also to jointly manage all simultaneous demands in order to guarantee the speed and robustness of the support system [ZGA 07a].

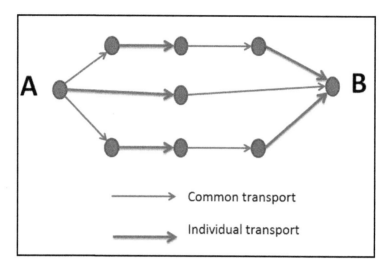

Figure 2.2. *Co-modality. For a color version of the figure, see www.iste.co.uk/zgaya/logistics.zip*

2.3.3. Chosen approaches

2.3.3.1. *Management system of shared vehicles integrating multi-agent coalitions for the optimization of co-modal itineraries*

In the context of this research, we have focused on the conception and implementation of a shared vehicle management system called SITCoMo[1], based on a multi-agent architecture and an optimization approach applied to a co-modal graph [JER 12]. The problem analyzed here covers all means of transport, including carpool (shared private transport), public transport, car share (free car sharing as in the autoLib mode), bike sharing (free bike sharing as in VLib), etc. The problem regarding optimized and dynamic car share is introduced and analyzed in *Sghaier's* thesis [SGH 11] and in the works of Ben Cheick [BEN 14a].

At the moment t, the problem of car share management described above is defined by:

– N queries formulated during a Δ_ε time interval. All these queries are denoted by I_t.

---

1 *Système d'Information de Transport Co-Modal* (Co-Modal Transport Information System).

– $I_k(d_k,a_k,W_k) \in I_t$ is an itinerary demand formulated by the user k at the moment t, from the departure point d_k toward the arrival point a_k during the time interval $W_k = [td_k, ta_k]$, with td_k and ta_k, respectively, corresponding to the earliest departure time (minimum departure time of d_k) and the latest arrival time (maximum arrival time of a_k) with $t \leq td_k \leq ta_k$.

– $R_g(d_g,a_g,W_g)$ is an identified route to respond to one or more queries belonging to I_t. A vehicle completes this route from the departure point d_g to the arrival point a_g, during the time interval $W_g = [td_g, ta_g]$, with td_g and ta_g, respectively, corresponding to the earliest time to leave (d_g) and the latest time to arrive at (a_g).

The architecture of SITCoMo is based on the interaction of six types of software agents (Figure 2.3): Interface Agents (IA), Super Agents (SupA), Transport Service Agents (TSA), Transport Information Agents (TIA), Router Agents (RA) and Evaluating Agents (EA):

– IA plays the role of interface between the user and the system. It assists the user in the creation of his query and transmits it to SupA;

– SupA are in charge of determining different routes than can respond to diverse queries in cooperation with TSA and TIA;

– each transport service in SITCoMo is represented by a TSA regardless of operators. For example, TSA_1 for car share, TSA_2 for public transport, etc.;

– each transport operator offering services for SITCoMo is represented by a TIA. A TSA is then responsible for TIA in the same service. For example, in France, the operators from public transport *RATP*[2] (Ile de France) and *Transpole* (Lille) are each represented by a TIA. These two agents are associated with the same TSA representing public transport. The dynamic carpool system proposed by *Sghaier* [SGH 11] is represented by a TIA associated with another TSA representing the carpool service;

– communication between agents (TSA and TIA) and the SupA allows the identification of all the possible routes to respond to different user queries. One itinerary can be done by various types of vehicles belonging to different means of transport. This is represented by an edge on a graph moving from one point to another in a precise time window. The nodes in the graph represent the connection points for changing the means of transport from one route to another;

---

2 *Régie Autonome des Transports Parisiens* (Autonomous Operator of Parisian Transports).

– SupA applies the first optimization approach on the graph composed by these routes. The results obtained from the algorithm execution constitute a graph, called the shortest path graph. A route of this graph exists if there is at least one means of transport offered by a supplier in a time window corresponding to the demands of the traveler at that time;

– each route in this graph is represented by a route agent RA, created by SupA. Each RA then applies an evolutionary approach in order to choose the best means of transport, according to the travelers' expectations, for example, and the number of available places. A RA coalition is finally elaborated in order to build different possible *route combinations* to respond to users' queries. In SITCoMo, a coalition of RA is a temporary group of agents which can build a complete itinerary. The connection points between these routes represent the starting and finishing points of each route;

– after coalitions take place, EA receive the different possible *route combinations* in order to calculate the best solution. This solution is global to all user queries received and managed in real time. Each user receives his answer through his own IA.

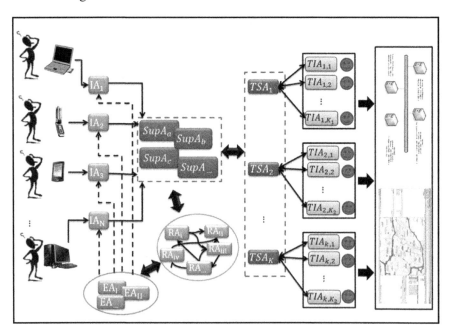

Figure 2.3. *SITCoMo multi-agent architecture for optimizing co-modal itineraries*

Optimization algorithm

As we have explained previously, each edge of the final graph is deduced after the calculation of the shortest path in terms of time. This is called $R_g\left(d_g,a_g,W_g\right)$ route and is represented by a RA. SupA creates as many RAs as the number of routes that will appear on the final graph.

For solving this optimization problem, a genetic algorithm has been developed [JER 12, ZGA 07a]. A solution for this proposed algorithm is represented by a matrix-shaped chromosome where lines represent all the users globally who compose simultaneous queries $\left(I_t\right)$ and columns represent different identified vehicles, represented as $V_h^{R_g}$ ($1\leq h\leq H$) to cover the $R_g\left(d_g,a_g,W_g\right)$ route. Each element in the matrix indicates the attribution of a $P_p\in P_t$ user to a $V_h^{R_g}$, $1\leq h\leq H$ vehicle, as follows:

$$CH[p,h] = \begin{cases} 1: & \text{If } P_p \text{ is attributed to } V_h \\ *: & \text{If } P_p \text{ can be attributed to à } V_h \\ X: & \text{If } P_p \text{ cannot be attributed to } V_h \end{cases}$$

Each user must be attributed only once to a single vehicle, respecting his preferences and constraints. For example, when a person will not drive a shared car, the system takes this constraint into account in the attribution process: we place an X in the case corresponding to the carpool on the attribution matrix (the chromosome).

Chromosome evaluation

A chromosome must be evaluated by three functions, because the proposed approach is to solve the multi-criteria optimization problem. The criteria taken into consideration are the journey's cost (in Euros), the total time for the trip (in minutes) and greenhouse gas emissions (in carbon equivalent grams).

In order to formulate the cost for each route, we use the binary variable $X_{V_h}^{R_g}$: transport of the person using the $V_h/1\leq h\leq H$ vehicle for the $R_g\left(d_g,a_g,W_g\right)$ route:

$$X_{V_h}^{R_g} = \begin{cases} 1, \text{if the } V_h \text{ vehicle is used for covering the } R_g \text{ route} \\ 0, \hspace{5cm} \text{if not} \end{cases}$$

We have given the following variables $C_{V_h^{R_g}}^1, C_{V_h^{R_g}}^2, C_{V_h^{R_g}}^3$ in order to, respectively, calculate the cost (in Euros), time (in minutes) and greenhouse emissions (in carbon equivalent grams), if the $V_h / 1 \leq h \leq H$ vehicle is used for covering the R_g route.

When $C_{R_g}^1, C_{R_g}^2, C_{R_g}^3$ are the different global costs for a $R_g(d_g, a_g, W_g)$ route, which corresponds to a CH chromosome, we then have:

$$C_{R_g}^1 = \sum_{h=1}^{H} C_{V_h^{R_g}}^1 . X_{V_h}^{R_g}$$

$$C_{R_g}^2 = \sum_{h=1}^{H} C_{V_h^{R_g}}^2 . X_{V_h}^{R_g}$$

$$C_{R_g}^3 = \sum_{h=1}^{H} C_{V_h^{R_g}}^3 . X_{V_h}^{R_g}$$

Coalition constitution for co-modal itineraries

For our system, we have proceeded to identify route combinations thanks to the forming of RA coalitions (section 1.7.1.2.2). In this context, agents are identified and temporarily gathered to solve a co-modal transport problem. This gathering (coalition) enables agents to satisfy the needs requiring the synergy of competencies from each group member.

Coalition formation process for RA

At time t, we consider a set A of n agents RA, $A = \{RA_1, RA_2, ..., RA_n\}$. The set $I_t = \{I_{1,t}, I_{2,t}, ..., I_{N,t}\}$ is proposed to these agents. The aim of each RA (represented by a matrix) is to take part in one or more coalitions in order to constitute optimized co-modal itineraries (Figure 2.4).

The process of coalition constitution is implemented, thanks to interaction [GEN 10, MÜL 06]. In this work, we have suggested interaction strategies enabling agents to form coalitions.

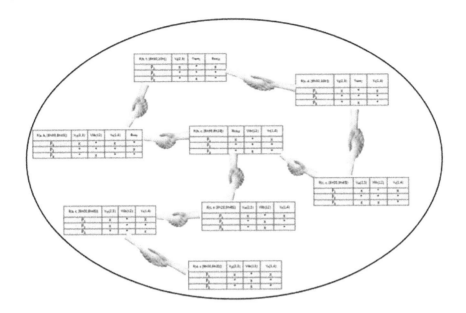

Figure 2.4. *Coalition formation of RA*

Knowing that a route is represented by an agent, each coalition forms a route combination that can be a possible solution for a user query. An agent can be the initiator and then ask other agents to form a coalition with him, as it can also receive invitations from other agents to form a coalition with them.

Making reference to the works of Müller *et al.* [MÜL 06], we consider that a RA can perform the following roles:

– *Candidate*: the RA still does not have any partner. He expects to take part in coalitions.

– *Initiator*: the RA decides to start a coalition, it sends requests to different agents to reply to a query.

– *Solicited*: the RA receives a coalition proposal on behalf of another agent.

– *Member*: the RA accepts to take part in a coalition initiated by another RA.

The aim of all RA is to find all the possible route combinations to reply to a set of user requests. They adopt the role of Candidate at the beginning of the process, waiting to start or to receive coalition proposals.

The coalition process starts by identifying initiating agents. A Candidate agent takes the role of Initiator if and only if his departure point corresponds to the departure point of at least one query (from the set of the simultaneous user queries). His departure time must be equal (with a small delay toleration margin) to that of one request.

In the cases where departure and arrival points for user requests are in a zone controlled by only one route agent, these queries can all be solved by this sole agent, making coalition strategies useless.

In the opposite case, the Initiator agent must invite all RA to form a coalition in order to satisfy user requests. In this case, the Initiator must send a coalition proposal to other RA Candidates. To optimize the number of messages sent, the Initiator demands SupA to make a list of the agents who could possibly be interested in forming this coalition. In fact, the SupA creates RA (according to the transfer graph for the calculated route) knowing that the SupA is the only agent that knows all the neighboring relations for each route. This fact will help avoid sending a large number of messages between all RA.

An agent which performs the role of "Candidate" may later become "Solicited", when it receives a coalition proposal from the part of an Initiator.

2.3.3.2. Dynamic carpooling service

In this section, we will focus on the problem of dynamic carpooling services, characterized by their flexibility in comparison with the classical model of carpooling, called "static". The majority of existing systems belong to this category. Often, they are non-automatic systems accessible through the Internet, integrating basic management functionalities for dealing with carpooling offers and demands.

Dynamic carpooling is more useful because it proposes a route solution in a relatively short time. Subject to previous registration, the route must be suggested almost immediately and can be modified in real time, according to travelers' new demands or even according to alert signals and risks of the transport network (traffic jams, breakdowns, etc.). Since the mechanism is connected in real time, such a service is possible thanks to new technologies (NICT[3]).

Implementing a dynamic carpooling system is a very specific research problem due to its real-time nature. In the context of the works of Sghaier [SGH 11], an optimized dynamic carpooling system has been proposed. This is called CODAC[4] (Figure 2.5).

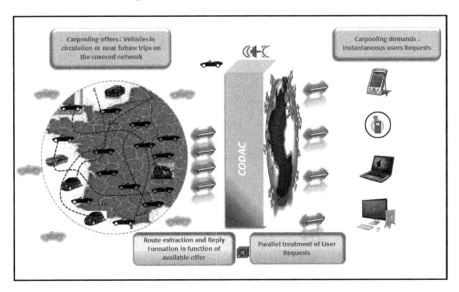

Figure 2.5. *Architecture of the CODAC system (dynamic carpooling)*

The CODAC engine is composed of optimizing agents which can optimize a complete route from the optimization of different parts of the circuit. The agents involved in optimizing the same circuit form a coalition, in which each agent is in charge of optimizing his circuit share and coordinating the strategy with other agents, what creates an optimized circuit for each traveler (Figure 2.6).

---

3 NT: New Information and Communication Technologies.
4 *Covoiturage Optimisé Dynamique basé sur les Agents Communicants* (Optimized Dynamic Carpooling System based on Communicating Agents).

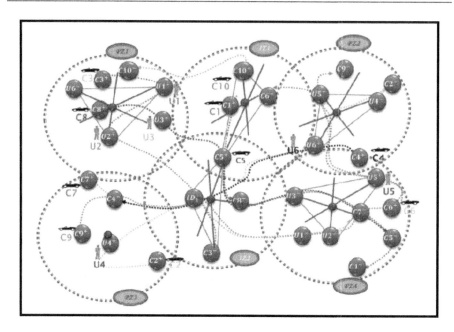

Figure 2.6. *CODAC: distributed optimization with communicating agents. For a color version of the figure, see www.iste.co.uk/zgaya/logistics.zip*

2.3.4. Outstanding results

In the transport field, the OSL team of CRISTAL is particularly interested in implementing systems that accompany the traveler in the most efficient and pleasant manner, before and during his journey to his destination. This company is manifested by the availability of services connected with transport, which include pertinent, interactive and instantaneous information. This information is placed in what we call RESAM[5]. For example, one service may correspond to an itinerary calculation in a multimodal transport network (for which the OSL team issued a patent back in 2006[6], the weather forecast, a cultural or touristic

---

5 *Réseau Etendu de Services d'Aide à la Mobilité* (Extended Network of Mobility Support Services).
6 European patent (August, 11th, 2006) under number 06291308.2 ref: 117.440, for the invention of "research procedure and itinerary composition" in the field of multimodal transport.

piece of information, a free parking space, etc.). Four theses have been produced, namely *Saad* [SAA 10a], *Sghaier* [SGH 11], *Jeribi* [JER 12] and *Bousselmi* [BOU 14], which contribute to the implementation of such a system.

The aim is to conceive, optimize and implement a Mobility Support Information Service System which can provide users with all the useful information before and during their trip (on foot, by car, etc.) on every kind of fixed or mobile device (smartphone, tablet, etc.), eventually attached to a circulating vehicle. The suggested interaction modes (intra and inter-systems) are based on high-level semantics which exceeds classical communication limits. This difficulty has encouraged us to conceive an extensive vocabulary library and specialized semantics based on a dynamic ontology dedicated to transport [SAA 10a]. Works presented in this section have been initiated in the context of the ANR VIATIC.MOBILITE project, from I-TRANS competitive pole: "Railway at the heart of innovating transport systems" (http://www.i-trans.org/index.htm), in which the OSL team is a full partner.

In addition, other partnerships have been initiated by this team in the context of the CISIT[7] project, financed by the CPER 2007–2013. Thanks to these partnerships, rich in research actions, the obtained results have been published in scientific works [HAM 12a, HAM 12b, SAA 10b, ZGA 09].

2.4. Crisis management logistics

2.4.1. *Context*

Research works presented in this section are the fruit of cooperation with the logistics department of the Airbus Group, formerly called EADS (European Aeronautic Defense and Space company). Crisis management consists of four complementary stages (Figure 2.7):

– crisis management prevention: reducing the risk of crisis emergence to the maximum;

---

7 http://www.cisit.org/, *Campus International sur la Sécurité et l'Intermodalité des Transports* (International Campus for Transport Security and Intermodality).

– the capacity for operational reaction: advanced strategic planning, training and simulation in order to ensure availability, mobilization speed and resource allocation to manage potential emergencies;

– declared crisis management: facing the crisis, while reducing its impact. It is necessary to find a solution in order to guarantee evacuation, search and rescue of victims, as well as securing crisis zones, minimizing its effects and limiting its impact on the environment and local population;

– analysis: analyzing the current situation, in order to reinforce the prevention system.

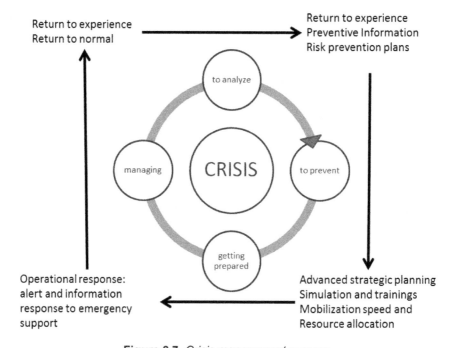

Figure 2.7. *Crisis management process*

2.4.2. *Problem*

These works aim at providing a solution with an innovative approach of organizational, technological and informational optimization for decision-making problems met in logistics information systems. The objective is anticipation, in order to prevent any stock rupture in the Crisis Management Logistics Chain (CMLC).

The studied CMLC is distributed, dynamic and composed of many zones. The choice of the zones location is a strategic decision of high importance. This decision is made after a crisis. These zones must conform to certain distance conditions, easy access and political stability in order to be valid and efficient. The stake is to satisfy each actor's needs in this chain in order to provide a better quality of logistics service. However, this chain is completely unpredictable. Its typology must be able to be modified in real time, in order to adapt to the real situation and to organize resource flow in the best possible way.

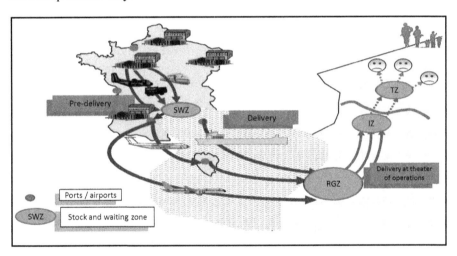

Figure 2.8. *CMLC zones*

The CMLC is composed of four main zones (Figure 2.8):

– SWZ (stock and waiting zone): the area where products must be gathered before shipping;

– RGZ (resource gathering zone): the base from which products are delivered to the crisis zone;

– IZ (intermediate zone): placed for strategic reasons. After satisfying its resource needs, the IZ transports the rest of the received resources from the RGZ to the terminal zone;

– TZ (terminal zone): the place where the victims of the crisis are waiting.

A supply chain like this will face delivery delays at every moment, erroneous consumption estimations, cargo losses, spontaneous peak consumptions and other unpredictable events. All these mishaps may provoke stock ruptures at any moment of the chain, which may result in dramatic consequences, to the point of human casualties.

These extreme situations are not acceptable in such a chain because of the human stakes, so it is necessary to create decision-support tools that will enable anticipation of certain disturbances, and, if necessary, simulation of the best strategies to adopt according to the critical situations.

In the case of humanitarian action, it is necessary to implement operations that will immediately include the development and maintenance of supply chains providing logistics function support. The reliability of the logistics chain demands that the chain's mission be successful in providing necessary resources to transfer points in the system in order to satisfy the demands of the emergency mission.

The problem of logistics in the context of humanitarian assistance is to deliver resources without delay or stock rupture that could disrupt the functioning of a zone. This is called supply policy and demands knowledge and total mastery of delivery deadlines and resource calculation.

The objectives of a CMLC are to:

– deploy rescue units, resources and the associated equipment (staff, vehicles, planes, etc.);

– rebuild infrastructure;

– supply water, food, clothes, etc.;

– provide medical support.

In order to achieve these goals, flow management and CMLC optimization are essential. Push and pull flows are particularly important (section 1.3.4). The principle corresponds to the restock demand made by the consumer depending on his needs and with regard to the margin fixed by order deadline. The anticipation of needs and the notion of security stock are necessary to avoid any breakdown or perturbation. This method ensures – at least in theory – permanent adequacy of support to real needs.

2.4.3. Chosen approaches

In a CMLC, actors are numerous and varied. Confronting the problem's complexity, we adopt A3C-2SL architecture described previously (section 1.9), in which the ground (third layer) corresponds here to a critical situation. Multiple modeling approaches are possible. Nevertheless, they all require the integration of all the zones of the supply chain, modeling them through one or more autonomous and interactive entities, which we call "zone-agents". The MAS corresponds to the second layer of A3C-2SL (Figure 2.9).

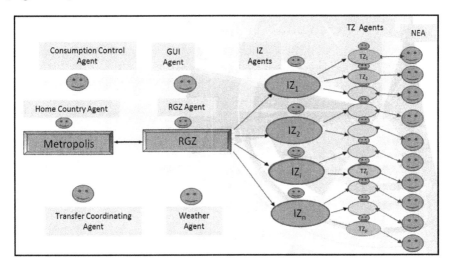

Figure 2.9. *Suggested multi-agent organization (second layer of A3C-2SL)*

In the second layer of A3C-2SL, agents constantly receive information from the lower level: theater of operations. The connection can be made with

RFID[8] technology: for example, a truck equipped with a RFID chip going through a portico can be detected and the information transmitted to agents. From this information and different mathematical models, agents adapt their behavior in order to better respond to the need of the lower level. This may correspond to correction and adjustment of the mathematical model.

2.4.3.1. Diffuse agent-based scheduling

In contrast to a non-diffuse resolution approach, where scheduling decisions are calculated by a central entity, we propose to solve the problem of resource allocation in each zone of the CMLC by a collaborative optimization approach, considering that decisional entities are decentralized in a distributed organization (Figure 2.10). The problem of global scheduling is then divided into many local scheduling problems. Each agent is associated with a zone and thus controls its own resources and possesses its own decisional autonomy.

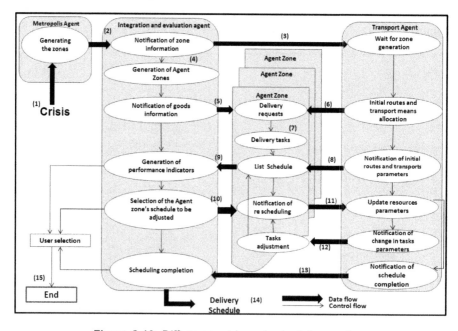

Figure 2.10. *Diffuse agent-based scheduling system*

---

8 Radio Frequency Identification.

2.4.3.2. OBAC demonstrator

The OBAC[9] demonstrator has been developed to implement the agent-based architecture described earlier in this chapter. This demonstrator is provided with multiple functionalities implied in the optimization process of the logistics chain and in the answer to the crisis. Among its functionalities, we find the positioning of logistics zones, the parallel treatment of resource delivery orders and the optimized estimation of needs: OBAC optimizes amounts, reducing stock shortages and ordering only the necessary quantities.

2.4.3.3. Implementation example of the OBAC demonstrator

The OBAC demonstrator enables us, for instance, to follow the stock evolution in logistics zones for water and supplies. First, for the SWZ (Figure 2.11), then for the IZ (Figure 2.12) and, finally, for the TZ (Figure 2.13 for TZ1 and Figure 2.14 for TZ2). In these figures (to the left of water and to the right of supplies), the ordered and consumed amounts are, respectively, represented in "green" and "blue".

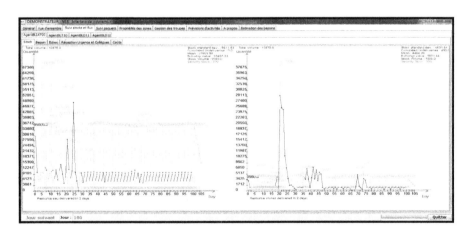

Figure 2.11. *Stock evolution of the SWZ. For a color version of the figure, see www.iste.co.uk/zgaya/logistics.zip*

---

9 *Optimisation à Base d'Agents Communicants des flux logistiques pour la gestion decrises* (agent-based optimization communicating logistical flows for crisis management).

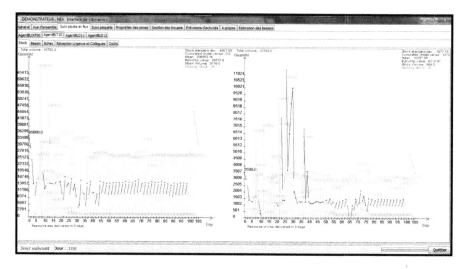

Figure 2.12. *Stock evolution of the IZ. For a color version of the figure, see www.iste.co.uk/zgaya/logistics.zip*

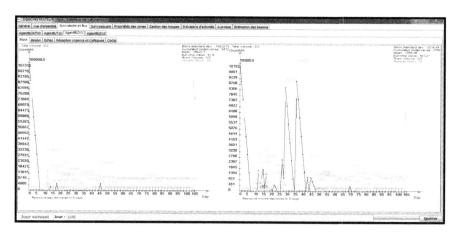

Figure 2.13. *Stock evolution of TZ1. For a color version of the figure, see www.iste.co.uk/zgaya/logistics.zip*

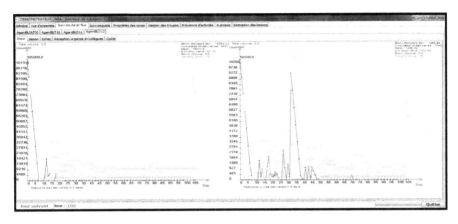

Figure 2.14. *Stock evolution of IZ2. For a color version of the figure, see www.iste.co.uk/zgaya/logistics.zip*

For the stock in IZ, OBAC cannot provide the necessary amounts to TZ, in order not to incur in stock shortages. It only orders the necessary amount in order to have enough resources until the end of the rescue (especially water supplies). The same happens with the SWZ, which chooses to order a large amount of resources in order to have enough resources for later use. Adding initial stocks and delivered amounts to SWZ, we obtain all the resources sent to the humanitarian aid operation. A comparison with real data is represented in the graph of Figure 2.15.

2.4.4. *Outstanding results*

In case of humanitarian crises in any region worldwide, a supply chain must be implemented to deliver goods from possession zones (home countries, like France, for example), to zones in need (victim zones). However, it is never easy to anticipate the evolution of a supply chain, nor the context in which this evolution will take place. Consequently, integrating disturbances as a parameter in the study of the chain can limit its vulnerability. The main purpose here is to foresee not only the amount of resources that will be consumed in a humanitarian aid supply chain, but also the interaction between different logistics zones in case of disturbance. Such an application must enable, in particular, the anticipation of logistics flows between the different zones of ground action and reduction of the impact of possible disturbances to these zones.

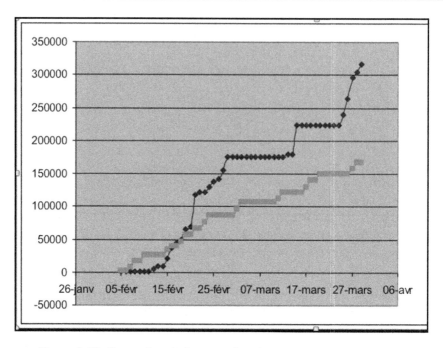

Figure 2.15. *Comparison between ordered amounts and sent amounts with OBAC. Ordered (cumulated) real amounts are shown in blue. Sent (cumulated) amounts with OBAC are shown in pink. For a color version of the figure, see www.iste.co.uk/zgaya/logistics.zip*

In this way, the results obtained in the works of Aida Kaddouci [KAD 12], Nesrine Zoghlami [NES 08] and Othman *et al.* [OTH 14] refer to the conception of a modeling and simulation approach oriented to the agents of supply chain in a context of strong disturbances. In fact, working in an uncertain environment encourages us to rely on DSSs in order to ensure the satisfaction of all the actors in the chain. Hence, the development of the OBAC demonstrator, which responds to the needs of a flow manager working on a distributed supply chain, as the humanitarian aid supply chain. In the context of a R&D partnership, the OSL team has guaranteed technological transfer, delivering operational support to the logistics department of Airbus Group, via the OBAC demonstrator.

2.5. Warehouse logistics

2.5.1. *Context*

The research works introduced here, "Increased reality at the service of warehouse logistics", were produced in the context of an industrial research project (2012–2015) between the OSL team and GENERIX group.

In this project, we are particularly interested in warehouse logistics, particularly in the essential stage of order preparation. In fact, ground experiences show that order preparation costs around half of the total expenditures in a warehouse [GUD 10]. This is due to the complexity of this function.

To facilitate and speed up order preparation, we use the NICT within the warehouse management systems. In this way, while vocal recognition is starting to be massively used in warehouses, another revolution is being announced: a guided system based on Augmented Reality (AR) technology. AR is a relatively recent technology that enables us to insert 3D images in a real environment. It can be defined as a real-time superimposed informatics frame, in 2D or 3D to the perception we have of reality. The purpose is not to create a virtual world, but rather to insert virtual elements in the real world, and to invite users to interact with these virtual objects. AR is integrated into our environment through visual mobile supports such as glasses, tablets and smartphones, in order to superimpose synthetic images to our perception of reality, in real time.

Until recently, AR was a concept relatively unknown to the wide audience because it lacked a support that could be accessible to all. The development and multiplication of mobile supports in recent years has encouraged a renaissance of this concept. Today, this new technology is used in different fields [CIR 13]; the most successful applications are in the fields of medicine [LEE 13], entertainment [MAC 02], art [GER 12], education and learning [VAT 12], industry [NEE 12] and aeronautical maintenance [HIN 11].

2.5.2. *Problem*

The challenge of this industrial research work is to improve logistics in warehouses by developing new systems based on cutting-edge technology (connected objects, augmented reality, etc.), by filling the gap in the existing systems. The objective is to optimize the operators displacements in warehouses, in order to improve their comfort and their productivity.

In fact, at present, warehouse logistics is managed by a WMS[10] type of software, integrating the structure's Logistics Information System (section 1.6.3). WMS centralizes orders and distributes them to packers. Currently, these operators receive their orders via a vocal system. At work, they carry headphones throughout the day and a synthetic voice dictates what they must do. They also have to confirm at a microphone each accomplished task. Even though vocal technology has revolutionized warehouse work thanks to its efficiency, it is highly criticized because it should be perfected from the point of view of productivity, preventing worker fatigue.

Thus, with the assistance of AR, the aim is to build an application prototype, to be used with AR "intelligent eyeglasses", in order to guide operators inside warehouses. These will soon be guided through AR technology, no longer vocal technology, because this appeals to the worker's visual organs. More interactive than vocal technology, these goggles enable the operator to visualize the needed information in order to prepare an order, while keeping his hands free.

Finally, despite the fact that AR has the potential of providing numerous benefits and becoming the dominant technology in warehouse logistics, to the point of completely replacing current technologies (such as *picking-by-voice*), the purpose is to develop, over time, an innovative solution which efficiently combines two complementary types of technology in order to achieve better results.

2.5.3. *Chosen approaches*

Figure 2.16 represents the global architecture of a solution called RASL[11]. This is composed of a WMS server which integrates an optimization agent-based program, which optimizes the operators' pathways in the warehouse. Article identification inside the warehouse is done with QR[12] codes, thanks to AR eyeglasses.

---

10 Warehouse Management System.
11 *Réalité Augmentée au Service de la Logistique* (Augmented Reality at the Service of Logistics).
12 Quick Response.

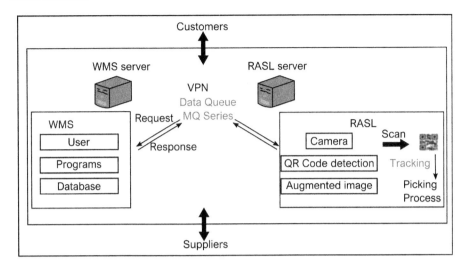

Figure 2.16. *RASL: proposed architecture of the AR system for warehouse management*

Again, in this case we have used the A3C-2SL model (section 1.9), but this time for optimizing warehouse management (hence, the third ground layer: warehouse). We have used an advanced algorithm for optimizing the operators' pathways as a function of the order flow coming from WMS (algorithm in the first layer of A3C-2SL). This algorithm is integrated into the behavior of the optimizing agent in the second layer of the A3C-2SL model. In the ground layer, each operator is guided via intelligent AR eyeglasses, according to the optimization algorithm.

2.5.4. *Outstanding results*

The *Generix Group* editor introduced the first functional prototype of RASL at the SITL 2014[13]. A press release was issued on March 24, 2014 in order to announce and present the product[14], the first AR application which contributes to the management of warehouse supply chains. Equipped with special goggles, the order packer picks the correct product from the right

13 *Semaine Internationale du Transport et de la Logistique* (International Week of Transport and Logistics) (April 1–4, 2014).
14 http://www.generixgroup.com/fr/actualites/Generix-tv/11953,lunettes-connectees-entrepot-Oscaro.htm.

place, while keeping his hands free (Figure 2.17), following the optimization algorithm developed in the first layer of the A3C-2SL architecture (section 1.9). During the transport week SITL 2014, *Generix Group* was awarded with the prize "best innovation in logistics products", for its application of connected eyeglasses in warehouse[15], resulting from the collaboration with the OSL/CRISTAL team from the *Ecole Centrale de Lille*.

Figure 2.17. *Warehouse order management: AR eyeglasses solution*

These tasks have been transposed to the health field, in particular, to the preparation of medical orders at the neonatal department of the *CHRU de Lille*. The nurses of this service tested the eyeglasses, in order to secure complex drugs' preparations, requiring great vigilance.

2.6. Conclusion

In this chapter, we have introduced the research works of the OSL/CRISTAL team in the context of collaborative optimization at the service of logistics systems in the fields of transport, crisis management and warehouse management. We have focused in particular on the most astonishing results, proving the efficiency of the alliance between MAS and optimization. These works have been at the heart of a deep study for the resolution of logistics problems in the field of health. In fact, logistics problems in the health environment are practically the same as in industry

15 http://www.generixgroup.com/fr/actualites/communiques/12050,prix-meilleur-produit-logistique.htm.

nowadays, because a hospital has to be managed as an enterprise, while valuing care actions and being competitive at the same time. This requires optimized logistics management (methods and tools) with innovative modeling and optimization techniques in health business processes, that is, every process that involves improving patient care in health centers. However, and following the analogy with logistics problems in industry, the healthcare service does not concern "objects", but human beings whose desires and needs must be taken into consideration.

Thus, in the field of health, the aim of logistics is to find the most efficient way to take care of patients, not only in healthcare centers (internal logistics), but also throughout the supply chain. The patient pathway can be assimilated into a production chain with different constraints, such as time management or the necessary skills from the medical staff, to perform healthcare. The purpose is to find the best possible organization depending on the patients' needs, by efficiently distributing material and human resources in terms of availability and skills.

3

Health Logistics: Toward Collaborative Approaches and Tools

3.1. Introduction

In the field of health we traditionally speak of health logistics, which is split into two sections: hotel services (meals, laundry, etc.) and medico-technical services (which strictly refer to medical activities).

Logistics in the field of health developed with demographic, socioeconomic and regulatory evolution [BEN 12]: new organizations, new evaluation modes, etc. This new context encouraged health centers to adopt a more rigorous and objective cost-control policy, while ensuring the quality and security of the care provided to the patient. The purpose is to use the means allocated to the health sector efficiently, without excess, optimizing care for the patients. This optimization process can only be carried out by considering the overall healthcare system (not only hospitals). Hence the need for a more recent and general expression in this field: health logistics (HL), which includes hospital logistics (Figure 3.1).

The conception of healthcare becomes more global, with a patient-oriented vision, providing the patient with all the favorable conditions for a quality service. These favorable conditions equally concern the medical staff, who represent the ultimate human resource for care production. Favorable conditions for patients and for medical staff have become a major stake. Thus, health centers wish to improve performance, care quality and working environment. The optimal solution consists of implementing a

Chapter written by Hayfa Zgaya, Slim Hammadi and Jean-Marie Renard.

new functioning mode based on the expectations of patients and staff, establishing a diagnosis of current organization modes and adopting new tools and approaches. In this way, logistics is at the service of human beings, not the other way around.

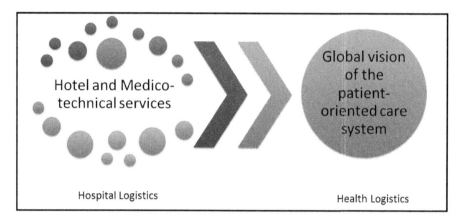

Hotel and Medico-technical services

Global vision of the patient-oriented care system

Hospital Logistics

Health Logistics

Figure 3.1. *Evolution of health logistics*

The objective of health logistics is to optimize patient care, focusing on quality, by reducing waiting time, limiting medical errors, etc. With the evolution of logistics, classical methods and tools are no longer enough. Now, the purpose is to adopt new information and communication technologies (NICT) in order to better exploit innovative approaches and tools.

In this context, the success that industrial logistics has met in the world of enterprises today encourages healthcare professionals to import this expertise and adapt it to the healthcare system, in order to achieve the objectives indicated above. Hospital restructuring has led a great number of hospitals to reconsider their operational processes, and to engage in modernization projects of infrastructure and organization.

In fact, the different nodes of the care network are complex systems which demand clever solutions enabling appropriate and dynamic resource allocation (human and material). Problems in this field are no longer limited to the search of funding sources. The main objective today is to find new process-oriented management methods. Management methods thus become reasoning methods based on logistics [KAL 04]. Besides, in comparison with other fields, health logistics is essentially characterized by its complexity, because a significant number of laws and regulations must be respected [DOE 07].

3.2. The health sector

3.2.1. *International context*

Hospital and healthcare systems in the majority of developed countries encounter difficulties associated with the important evolution of their socio-economic environment and their structures, as well as the increase in constraints and their missions [NAT 08, PIN 11, SCH 12]. Even though these systems may differ because of their objectives, their organization and their functioning principles, the problems with which they are confronted are similar: new diseases, rapid evolution of medical technologies, aging populations and budget constraints.

In the United States, even with a budget equivalent to almost 16% of its gross domestic product (GDP), it is widely known that, today, their health system does not provide efficient care and suffers due to the number of medical errors and useless interventions [GAL 09]. [WIT 12] recalls in a recent report about President Obama's project for reforming the American healthcare system: *The healthcare system in force in the United States arouses a number of questions in Europe. It is one of the most complex systems in all of the OCDE[1] countries, but certainly not one of the most reliable ones, neither in terms of public health nor in terms of service quality or accessibility.* The American Academy of Sciences and the American Academy of Medicine warn against the over-use or the under-use of resources, activity duplication, and systemic failures in the production of care and organization inefficiency [REI 05].

Beginning in the 1990s, the majority of the countries in Europe have decided to reform their healthcare systems [SAL 97]. These reforms feature the progressive adoption of a transnational and interdisciplinary perspective in order to improve the performance of healthcare systems [MCK 02]. In most cases, debates concerning the reform of the health system essentially refer to the right and access to care, as well as the cost, organization and

---

1 www.oecd.org/fr/els/systemes-sante/Panorama-de-la-sante-2013.pdf.

quality of this care. For example, a study from the National Research Council [STE 09] estimates that the explanatory factors for this situation are:

– the bad organization of care;

– an intermittent, complicated, unplanned and stressful patient pathway;

– the multiplicity of organizations and of health actors, which induces lack of clarity in the payment modes;

– insufficient healthcare centers and lack of integrated NICTs.

The majority of studies particularly recommend not only strong cooperation between healthcare professionals, researchers and firms in the field of system engineering, but also the use of methods and technologies which have proved to be efficient in other fields (industry, telecommunications, etc.). In fact, these technologies present strong potential for improving the health system at all its levels (patient, care team, organization and environment). These recommendations suggest a fundamental point dealing with a global procedure which integrates the use of NICT.

This confirmation agrees with the position of the European Union about the challenges in the field of health, in accordance with the WHO which encourages establishments and care organizations to engage in a process of reorganization regarding its different missions: patient care, research, teaching, logistics and human resource management. The impact of NICT is equally acknowledged [SCH 13]. These observations are also visible in the study carried out by the European Observatory on Health Care System [MCK 02].

3.2.2. The French context

The 2013 edition of OCDE's Health report[2] states that, according to indicators of health expenses, today France is known in Europe for its healthcare expenses, devoting 11.6% of its GDP to health expenditures. Life expectancy is higher than the average of OCDE countries, and the cover rate by the general Social Security regime is 77%, one of the highest in developed countries. However, beginning in 2000, numerous reports and

---

2 *Panorama de santé de l'OCDE*, édition 2013.

studies have described a picture of the state of the French hospital and healthcare system, which is – according to the authors – in moral, demographic and financial crisis [COU 03a, MOL 05]. In their book [MAR 08] have summarized the causal factors for this situation and raised the question of the performance of hospital systems, by relying on the leading principles of industrial engineering.

Indeed, this situation results in the progressive accumulation of new constraints associated with social, demographic and technical evolutions, combined with strong structural rigidity. Thus, hospital systems, emergency centers and healthcare networks encounter more and more obstacles against realizing their missions.

In recent years, different financial and methodological solutions have been provided to respond to this problem: modernization plans (*Hôpital 2007 et 2012*). The plan *Hôpital 2007* had a triple objective: to alleviate the collapse of healthcare centers, to combine funding with activity and to improve internal functioning through a better dialog between doctors and directors. The plan *Hôpital 2012* is an extension of the 2007 plan. It aims to achieve four objectives: to finish the normalization of establishments, to update hospital information systems, to continue hospital restructuring and to implement the Schémas Régionaux d'Organisation Sanitaires[3] (SROS).

Many methodological contributions have equally been elaborated, and, in order to accompany this transformation, the French State has implemented a network of agencies and organizations that offer (to those establishments aiming for reorganization) methodological guidelines and special advice, whether thematic or functional. We can quote the example of the *ANAES*[4], which later became *HAS*[5] (in 2005) and proposed many methodological lists, with particular recommendations and methodological documents. The Fédération Hospitalière de France (FHF) has proposed many guides, and the Groupement pour la Modernisation du Système d'Information Hospitalier (GMSIH) has published many documents and guides related to hospital

---

3 Regional Diagrams of Sanitary Organizations (RDSO).

4 *Agence Nationale d'Accréditation et d'Evaluation en Santé* (National Agency for Health Credentials and Evaluation).

5 *Haute Autorité de Santé* (High Health Authority).

information systems, the control of health establishments and the execution of change.

More recently, reports have suggested a mutation of the hospital, of its governance, and more globally, of the French healthcare system [MIN 08, VAL 08]. These proposals were given concrete form by a law project [LEG 09].

Hospital and health sectors are thus confronted with new challenges and will continue to do so with an increasing intensity in the years to come (governance, control, organization, intra- and inter-hospital cooperation, resource optimization, evaluation of performances, etc.).

Maintaining the quality and accessibility of all care services constitutes a key element in long-term growth and development, and constitutes a major preoccupation. However, the actors from the hospital and healthcare sector are relatively under-prepared and not specifically trained for solving the new problems by which they are and will be confronted. These actors do not have decision support tools available to them which are adapted to the demands of future functioning modes in the hospital and healthcare system.

3.2.3. *The regional context (north of France)*

The problems discussed above naturally affect the region of Nord-Pas-de-Calais, making a social situation which is already delicate even more serious.

Despite significant efforts (an amount of 76.5 million Euros in the context of the PRSP[6]) and a net improvement on health issues, the region of Nord-Pas-de-Calais is still below standard in comparison with the French average financial and sanitary indicators[7]. For instance, the per capita allocation is one of the weakest in France and the region of Nord-Pas-de-Calais is under-financed in comparison with other regions. While the health demand is heavier in this region than in other regions, the human and financial resources are weaker.

---

6 Plan Régional de Santé Publique 2007–2011 (Regional Public Health Plan 2007–2011).

7 "The condition of health in the region of Nord-Pas-de-Calais", *La Voix du Nord*, March 2007.

However, this region has record values in practically all pathologies in comparison with the national average: cancer (breast: +32%, aero-digestive tract: +81% and lung: +22%), cerebrovascular disease (+15%) and cardiovascular disease (+50%). Mortality rates are among the highest in France, with a particularly significant rate in patients below 65 years of age.

The access to quality healthcare remains uneven over the national territory, and social or regional disparities in terms of life expectancy are worrying: 9 years' difference in life expectancy at 35 years of age, between a skilled and an unskilled worker, and 5 years' difference in life expectancy between a man living in Nord-Pas-de-Calais and a man living in Île-de-France.

The medical services on offer are not adapted to the regional situation. General practitioners are unevenly distributed and there is a strong lack of specialists (–30% in comparison with the national average) in numerous fields, especially ophthalmology and pediatrics. As a consequence, delays to obtain an appointment may last up to many months, and that is the reason why patients resort to the emergency services, often saturating them.

However, we can observe that the implementation of the Regional Public Health Plan 2007–2011 promises an appreciable mid-term improvement of the situation (particularly in hospital equipment), due to the strong-willed policy of the region.

We equally observe that there is an increasing awareness about the major stake which represents the global organization of healthcare and patient circuits. Control is conditioned by different facts: not only the realization that hospital and healthcare are complex systems, but also the need to efficiently manage health costs, risks and care quality. To master all the aspects of its dynamics constitutes a major problem. That is why attaining the necessary level of performance is a challenge demanding a global answer.

To achieve the evolution of hospitals and healthcare systems, we must respond to multiple needs. To implement these much needed reforms, establishments and healthcare organizations must respond to real expectations with innovative methods and tools at all levels: strategic, tactical and operational.

3.3. Emergence of new needs

The analysis of the dysfunctions of hospital and healthcare systems, particularly those observed in hospital services and care centers, shows that these problems stem from a bad adjustment to constraints and to the evolution of their missions, as well as to a bad management of patient flow [GEN 13]. Thus, optimizing the organization and its information systems constitute important levers for the success of this evolution. Generally speaking, the implementation of strategic, tactical and operational management of care production systems is now indispensable.

To make this transformation successful, it is imperative to define new organizational paradigms, and new professions for the management and supervision of these new organizations as well as their support and appropriation mechanisms.

The reconfiguration and improvement of healthcare systems need a "blast reorganization". The complexity of the problems calls for an innovative, global and scientific approach. Therefore, responding to this objective demands an active approach to health logistics.

3.4. Health logistics

Hospitals and healthcare systems constitute particularly complex socio-technical organizations. This complexity can be explained not only by the technical nature and variability of deeds, but equally by the specificities of medical activity and patient–physician relationships. These systems are subjected to unprecedented mutations. In fact, regardless of their dimension or their missions, hospital systems have had to face an increasing demand for care, user behavior changes, set of reforms: *Plan Hôpital 2007* and *2012*, new rates for T2A[8] activities and new governance systems.

For all health professionals (doctors, paramedics, administrative employees and technicians), this mutation presents important risk factors associated, among others, to the loss of reference points, the questioning of job description reference files and the introduction of new functions. Indeed, these professionals are confronted with problems of increasing complexity,

---

8 NT: "*T2A*" stands for *Tarification a l'activité*, which is a funding mode resulting from the *Plan Hopital 2007* reforms.

such as: How should patients' pathways in the context of care at home be improved, secured and optimized? How should the care of patients in distributed systems (care networks, global supply chain for emergencies) be improved? How should we combine humaneness with profitability, medical skill and productivity? Which NICTs to use and how to implement them in harmony with the specificities of hospital activity? Finally, which will be the new organizations that will respond to new sustainability restrictions?

Consequently, answers to these challenges are to be found through innovative, global and scientific hospital logistics, the only method capable of responding to these problems.

3.4.1. *Intra- and inter-hospital logistics*

Logistics, transversal activity, is a major function in the management of flow of hospital and healthcare systems (patients, products, equipment). It aims at the identification and optimization of different flows. We distinguish:

– nearby inter-hospital cooperative logistics: the organization of cooperation between distinct structures, namely mixed hospital campuses, "public/private" and laboratory/medical imaging subcontracting;

– intra-hospital logistics: the organization (or reorganization) of essential internal flows, as in the case of the extension of a building or the construction of a new building.

3.4.1.1. *Intra- and inter-hospital planning*

Planning must enable the coordination and coherence of the activities performed in hospital and healthcare systems, globally considering these activities at different levels:

– inter-hospital planning at a regional scale: sharing regional human resource pools between different entities, management of vacant beds and crisis management (as in the "white plan");

– nearby inter-hospital planning: planning of shared modalities between public hospitals, sharing pools of local human resources between many entities and planning shared modalities on mixed campuses (public/private);

– intra-hospital planning: medical time management, block planning and shared resources in hospitals as in re-education centers.

3.4.1.2. *Quality management*

The famous report published in 2001 by the Institute of Medicine [IOM 01] defines patient-oriented care as one of the main features of quality. From this perspective, hospital organization is increasingly called upon to adhere to the logics of organizational innovation, based on the continuous improvement of service quality and the cost reduction of functioning/stays, respecting the increasing and pressing demands of political, social and economic partners [SUF 12].

The 1990s witnessed the development of quality procedures in health establishments, under the form of programs based on regulatory (ISO 9000) and professional (HAS in France and JCAH in the United States) references, in the context of care supply regulation. However, the concept of innovation in hospital organization is often focused on the diversification of care supply, to the detriment of an organizational deficit. In this way, the organization has rarely integrated the innovation which could demolish the compartmentalization that governs its structures. Nevertheless, since the OCDE launched a project about quality indicators[9] for healthcare establishments in 2001 (in partnership with leading organizations and countries in the field of health), numerous advances have been registered, particularly thanks to the implementation of a conceptual frame and a methodological basis providing the necessary information about quality.

3.4.1.3. *Performance evaluation*

The evaluation of performances of the healthcare system must rely on scientific methods and tools, which will enable:

– evaluation of the health activity, aiming at content indexation and documentary and bibliographical management, applied to the normalization of indexes and semantic and thematic websites;

– formalization of medical knowledge for elaborating ontologies, case databases, knowledge databases, thesauruses and good practice guidelines.

---

9 Health Care Quality Indicators (HCQI).

These will be exploited in order to model recommendations following the guidelines of good clinical practice. They will also help to produce robust thesauruses (without coding support bias), conceive prevention and alerting systems, and optimize the patient pathway.

A volunteering policy centered on Hospital Logistics is particularly necessary for responding to the needs of healthcare system actors and for helping them to accomplish the necessary evolution of the health system.

3.4.2. Challenges

In the current regulatory context, health establishments must record their activities in a qualitative evaluation follow-up in order to obtain the V2 certification issued by the HAS. This evaluation refers to the quality of patient care and the security conditions of care activities. It is obligatory and conditions the activities of healthcare establishments. These establishments must also enforce the $PMSI^{10}$, which makes it possible to analyze their activities and costs. To this, we should add the rapidly evolving context of medical practices as well as the growth and diversity of care demand. Facing these constraints, hospitals are faced with many challenges. As in every industrial procedure, it is convenient to favor the "cost/quality/delay" trio. Trying to find the best compromise between these three criteria, even when referring to patient care, the quality criterion dominates the other two because of its tight link with patients' security [ETI 98, SAM 12]. As a consequence, the changes witnessed by hospitals nowadays demand not only a mutation in patient care but also an evolution in the economic control and management of certain care services. In the following section, we describe three key features of this mutation.

3.4.2.1. Control of health expenses

The French healthcare system is becoming more and more expensive, and we observe a constant increase in the allocated sums for health expenses. These expenses are essentially attributed to the Consumption of Medical Care and Goods[11]. According to INSEE estimations based on the work of

---

10 *Programme de Médicalisation des Systèmes d'Information* (Program of Medical Information Systems).

11 *Consommation de Soins et Biens Médicaux* (CSBM).

DREES in 2008[12], approximately 170.5 billion Euros were devoted to the funding of CSBM.

The important and increasing funding of the French economy makes it imperative to implement a more efficient organization. This will have to guarantee more rigorous resource management in order to control costs, but at the same time prove an improvement in care quality. This expense control constitutes a major challenge, because it greatly relies on the control of care production modes and organizational choices for health systems.

3.4.2.2. Care quality improvement

The evaluation of practices, procedures and results of care actions is at the center of the new health policy. It involves all dimensions: accessibility, globality, continuity, usefulness, medico-economic efficiency and even sociological evaluation. Since 1996, all public or private health establishments must enroll in a certification process. Accreditation is an external evaluation process for care centers, performed by independent consultants, and concerns all the functioning establishments and practices, as well as those from their tutelage organisms. It aims at ensuring that security and care quality conditions are met. In this way, health establishments are under the obligation of providing a quality service, at the best cost.

This quality is measured particularly according to the regulations that the hospital must follow, in terms of materials, processes and staff qualification. Quality is equally defined as the aptitude to satisfy expressed or implicit needs, by the engagement of the hospital structure and of professionals in permanent deeds and systematic improvement measures offered to the patient.

In a health establishment, measuring quality consists of regularly verifying the conformity of the organization to care, reception and others, in terms of previously defined performance levels. The quality policies which involve all the professionals within a care structure constitute a major stake of public health, in the measure that care quality improvement is done for the benefit of the patient. It also constitutes an organizational challenge because the management dysfunctions detected by the self-evaluation must offer the opportunity to rethink organizations. It is a real financial challenge, because the quality approach helps to avoid expensive dysfunctions.

---

12 http://www.drees.sante.gouv.fr/etudes-et-resultats,678.html?publication=2008.

Ultimately, quality can be improved in a significant way adopting an integrated strategy and reflecting the means and the resources deployed by different types of actors and services [TLA 09].

3.4.2.3. Resource optimization

The health sector is facing the evolution of a demand for care services, which is mainly determined by the fact that the population is aging; the improvement in the quality of life provokes an increase in care demand and entitles the patient to be more demanding in terms of care quality and its implementation [MAZ 10]. The care offered must follow the evolution of morbidity and the demands associated with it. The developments of ambulatory services, hospitalization alternatives, city-hospital networks and the deployment of new medical techniques illustrate this adaptation of the care offered. As a consequence, the hospital is today confronted with a rarity of certain resources which have become too expensive or unavailable. This has provoked a need for optimizing resources at health establishments, and also to adopt management tools for efficient resource allocation at the emergency services.

3.5. Hospital emergency services

Today, hospital emergency services (HES) occupy a strategic place in modern care systems and represent the hospitals' main responsibility [BEL 03, KAD 13]. This key role is to be reinforced in the future due to the sustained growth of patients demanding the emergency services in recent decades [COO 04]. These changes bring numerous problems to the different actors in public health, particularly functioning problems to face the rise of consultations and the cost of health expenses [COO 04, DER 00, NAT 08, PIN 11, SCH 12, SUN 13]. On the other hand, while it is true that the majority of care systems worldwide are confronted by this reality, the modes of facing it differ from country to country, in the measure that the care system is often a fruit of its history in each country [GAL 09, WIT 12]. Thus, in order to better understand current problems and stakes, it is important to first place emergency services in their general context, in order to better apprehend their functioning and typical features.

In this section, we highlight the specificities of the French care system and the consequences of the organization and general functioning of HES, in particular of pediatric emergency services (PES). In fact, this question is still relevant to France and hospital emergencies are considered to be at the highest standard in national organizations. DAREES launched a national investigation in 2013 concerning hospital emergency structures. This survey was designed to describe the pathways of patients resorting to hospital emergencies and the possible difficulties that may arise during service. In its 2014 report[13] about Social Security, the *Cours des Comptes* devotes an entire chapter to hospital emergencies, highlighting their attendance and the need to rethink the articulation of these typical structures with the city's medicine.

3.5.1. *The place of HES in the French health system*

The *Code de la Santé Publique*[14], more precisely, Decree No. 95.647-48 from May 9, 1995, relative to the reception and treatment of emergencies in health centers, describes the service missions, which must include:

– reception service: *to guarantee a 24-hour 365-day reception for every person involved in an emergency (including psychiatric emergencies) and to take care of the patient, particularly in a case of vital gravity*;

– definition of the demand and description of the patient's needs through symptom analysis;

– preserving the patient's life, stabilizing him with special treatment and *diagnostic gestures and/or emergency therapy* adapted to the situation;

– orienting the patient in the right direction, at the right moment, to the pertinent services at the right department;

– in certain cases, short-term hospitalization, according to the resources, plan and capacity of the establishment.

The mission of emergency services is incorporated within an evolving regulatory framework, which not only defines missions in terms of technical conditions for its functioning, but also fixes the threshold of activities. The size and frequency of changes in the regulatory framework reflect a permanent search for adaptation and the subordination of emergency services to regulations in force.

---

13 https://www.ccomptes.fr/Publications/Publications/Les-finances-publiques-locales2.
14 Public Health Code.

3.5.2. *Emergency services: a pivotal link and an alarming situation*

The nursing of a patient at an emergency service mobilizes different resources from health organizations: bottom-up at the hospital (the liberal sector, SAMU SMUR), at the care center itself (not only emergency structures but also administrative, clinical and medico-technical services) and top-down to care centers (other establishments). However, the notion of emergency in the medical field remains a source of great ambiguity. In fact, it is subjective from the patient's point of view, but not from the professional's perspective.

A review of the literature concerning the notion of emergency is presented in Table 3.1, which describes, for information purposes only, certain definitions provided by different authors/organisms. To define what can qualify as "emergency" remains a complex task [SOC 08a]. According to [AUB 03], it is a notion with entangled twists and turns, which manifests the relations between Man and Time. These relations have always been complex, evasive and turbulent.

Definitions	Sources
An emergency is "every circumstance which, after its happening or discovery, introduces or implies the suspicion of a functional or vital risk if a medical action is not immediately taken."	[SOC 01]
Vital emergencies must benefit from the most immediate care, be in the care services, research area or any other publicly accessible sector at an establishment.	[SOC 04]
True emergencies, conducing to a punctual and well-codified emergency medicine, risks hiding perceived emergency, that is to say, a real problem which must also be taken into consideration.	[ALV 05]
Social emergencies make part of the public emergency service.	[SOC 08a]

Table 3.1. *Definitions for the term "emergency"*

The waiting time and priority order at an emergency service are organized according to the degree of criticality of the emergency, with possible distinct waiting line zones. Many waiting lines of various types may coexist: a "standing patients" waiting zone, a "bed patients" waiting zone and eventually distinct zones as a function of specialized sectors. This emergency degree (Table 3.2) is represented by the Clinical Classification of Emergency Diseases[15].

Classification	Definition	Orientation
CCMU1	Lesion or vital forecast judged stable and abstention from complementary diagnostic deeds, or further at emergencies.	Functional unit
CCMU2	Lesion or vital forecast judged stable and decision for complementary diagnostic deed or further therapy at emergencies.	
CCMU3	Lesion or functional forecast judged susceptible of being aggravated in the short term, but does not engage vital forecast or the decision for diagnostic deeds or further therapy at emergencies.	
CCMU4	Pathological situation engaging vital forecast and whose care does not demand reanimation maneuvers at emergencies.	Severe unit
CCMU5	Pathological situation engaging vital forecast and whose care demands reanimation maneuvers at emergencies.	

Table 3.2. *Clinical classification of emergency diseases*

When the patient is admitted at the service, the responsibility is shared between nurses and doctors. The patient's exit is done as a function of the orientation.

Beyond this brief description, the nursing of a patient will be more complex according to the CCMU level and the potential interventions of the multiple actors at the PES. In order to account for this complexity, it is necessary to implement a system integrating behaviors and roles of the different participants surrounding the patient throughout the hospitalization process.

All the definitions of emergency quoted above highlight that, regardless of the type of emergency or the gravity of the problem, it is essential to immediately intervene and to respond to every demand, even if the service has not been programmed and does not involve vital risk. Thus, be it medical,

---

15 CCMU: *Classification Clinique des Malades aux Urgences.*

vital or social, the character of what constitutes an emergency refers to three basic elements:

– a given situation;

– a judgment concerning the situation;

– an immediate action tending to remediate the situation.

Hence, a situation is qualified as an "emergency" because it has been judged urgent and calls for an urgent answer. In other words, an emergency is often subjective and closely linked to the system of values of the person who deems the situation urgent.

This means that there often exists an important gap between the patient's perception of an emergency (perceived emergency) and its medical definition, according to medical staff criteria (real emergency). And this gap may also differ from case to case [SOC 08b].

In short, the emergency structure has become a pivotal link of patient care in health systems [BEL 03]. Emergency structures have been affected by mutations that have involved hospital systems. Apart from these mutations, we have to add the regularly increasing patient flow, which provokes overcrowding and perturbations in the working of these structures [DER 00, KAD 13, SUN 13]. The major stakes of these changes are decreasing waiting time (judged excessive by users), preventing structure saturation and maintaining security and quality for patient care [MOR 11]. Improving the organization of emergency care systems (ruled by decrees from 1995 and 1997) has become a necessity in a context where patient care is considered as one of the greatest priorities in the organization of care systems.

3.5.3. *Persistence of overcrowding risks at emergency services*

Reviewing the literature [CAR 14, MOS 09] proves that despite an existing explicit agreement between researchers all over the world – concerning the scale and gravity of the increased frequentation of emergency systems and its consequences – the debate is still ongoing with regard to the definition of "overcrowding" and the proper measures to deal with it. In the French literature, we find that researchers employ terms such as *congestion, overpopulation* and *bottleneck* [MAZ 10], even *overcrowding* and *crisis* [KAD 14]. In the English literature, we basically find two terms shared by the community of researchers: *crowding* and *overcrowding* [MOS 09]. To summarize, no matter the terms used in the literatures, the situation manifests the consequences of a lack of balance in the interaction of care supply and demand at the emergency services. This imbalance systematically generates overcrowding in the different units of an emergency service. As a consequence, we shall use the term "overcrowding" to designate the consequences of this bottleneck. The literature does not provide a definition for "overcrowding" that can be unanimously accepted by researchers [FEE 07], and its meaning may vary from organization to organization, according to the context and the available information [HOO 08, ROW 06, TRZ 03]. The concept of overcrowding in a care process is defined as an imbalance between the flow of patients and the establishment's size (emergencies or hospital), which may provoke negative consequences. It is then crucial to understand the causes for this overcrowding in order to remedy it.

From this point of view, [ASP 03] have elaborated a conceptual model that makes it possible to understand causal factors all through the care process. Other researchers have been inspired by this model [HOO 08, MAZ 10] to identify the causes that the authors split into three categories:

– input-related problems (*input*);

– problems encountered during treatment at Emergency Services (*throughput*);

– output-related problems (*output*).

Based on a cause and effect diagram, [KAD 13] have identified the main causes of overcrowding at an emergency service (Figure 3.2).

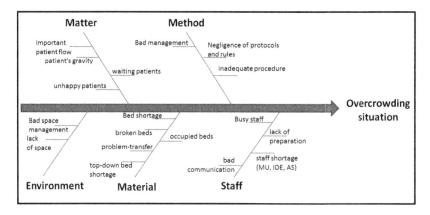

Figure 3.2. *Causes of overcrowding at emergency services, according to [KAD 13]*

The aim of this research is not only to understand the causes and consequences of bottlenecks at emergency services, but mainly to find a method for decongesting a saturated service. Generally, logic and reason demand that emergency patients be distributed where staff is more competent: at the department specialized in the pathology. For this, it is essential to remember that the classification into three groups of flows must not encourage us to compartmentalize care units at an emergency service. On the contrary, care emergency services need to be highly inter-operational and predictive in order to easily cope with patient flow. Inter-operability and anticipation applied to emergency services can help reduce costs and improve access to quality care. In concrete terms, if the patient's care at emergency services is an every-day problem, its improvement needs well-framed regulations. Among these rules, it is possible to quote a single rule that improve the patient care and then reduce the waiting time. This rule, to be applicable, primarily depends on anticipation. Communication between doctors and paramedics, and particularly administrative services, must be efficient in order to favor such anticipation.

3.5.4. Overcrowding at the French emergency services

In France, in the chapter devoted to hospital emergencies, the report from the *Cour des Comptes*[16] affirms that, despite the means deployed in the context of 2004–2008 emergency plans, difficulties are still witnessed in the functioning of these services. These difficulties (Figure 3.3) *result less from a lack of financial means, than from the capability of establishments to improve their internal organization.*

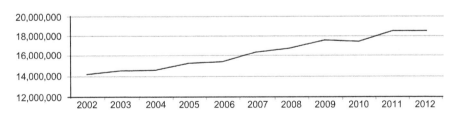

Figure 3.3. *Evolution in the number of patients at emergency services*

---

16 Sécurité sociale 2014 – September 2014. Cour des Comptes – www.ccomptes.fr – @Courdescomptes.

Moreover, the results from a survey conducted in 2013 by the Health Ministry in collaboration with regional agencies reveal that 100 establishments, mainly hospital centers but also university hospitals, had been identified by resource agent as being "in overcrowding" or "at overcrowding risk", with strong regional variations (Figure 3.4).

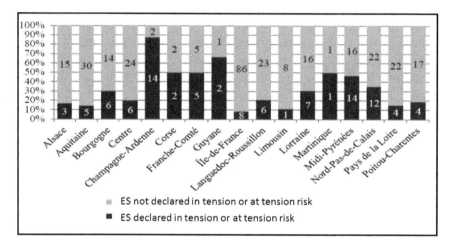

Figure 3.4. *Number of emergency services declared "in tension" or "at tension risk" regarding the total numbers of services, November 2013 (sources: [RAP 14])*

3.5.5. *PES: reality and challenges*

In the last decade, many measures were taken to restructure HES. They concerned both adult and child emergencies. As recalled by Hue *et al.* [HUE 08], numerous reforms in recent decades have contributed to deeply modifying the organization and emergency structural means, particularly those of pediatric emergencies. In the mid-1990s, these started to arouse an increasing interest in researchers, particularly in France and in Europe.

3.5.5.1. *Problems and challenges at PES*

In France, the PES have an established reputation and are integrated in the network of emergency departments (regulated on March 30, 2006) and the regional organization of children and teenage health services (regulated on October 20, 2007). In fact, the implementation of a structure exclusively devoted to pediatric emergencies (PES) relies on the need for these children to be welcomed at specific venues by a qualified medical and paramedical staff, in possession of adequate material adapted to children's needs. The younger the infant, the more serious the need for especially adapted means in order to treat these children and their families. This need is evidenced by the increasing number of children treated at HES. In the face of a sustained rise in the number of pediatric emergencies, numerous studies have been carried out since the 1990s to better understand the causes and the consequences of this phenomenon, in order to improve the organization of PES.

In recent years in France [DEV 97, STA 04, BRE 13], as in Europe or elsewhere [ELL 15, WHI 14, NAT 08], the demand for urgent pediatric care is increasing to the point of collapsing the entire care system (pediatric hospital poles, liberal pediatrics and infant care systems). Certain studies consider inflation, while others highlight the problem of false emergencies or "perceived emergencies".

In the majority of studies, criticism is always placed on the heavy burden of pediatric emergency consultations, estimating that these represent between 25 and 30% of hospital emergencies at the national scale [BRE 13]. Even if the majority of these consultations is done at liberal practices, a significant number of patients resort to HES, provoking saturation and problems specific to all the accompanying structures.

Today, the issue is relevant to France. According to [ARM 02], *while the child population's health condition has considerably improved, significantly reducing the number of hospitalizations and the duration of hospital stays, the resort to emergencies is exploding.* As highlighted by [BRE 13], the same remark is confirmed by the actors of pediatric emergencies: a rise in the frequentation of emergency services is observed everywhere. The increasing number of visits to pediatric emergencies engenders numerous problems, particularly for absorbing consultation flow and a non-negligible cost in health expenses. Babies under 1 year of age visit the emergency services three times more often than the general population. In the same way, 48% of parents resort to an emergency center at least once during the first year of life of their child.

The conclusion is that the situation of pediatrics in emergency services is considerable, complex and expensive. It is considerable in the measure that between 25 and 30% of emergency structures at the national level concern children. It is complex because the care process involves three actors: a medical team, a child and his parents [BRE 13]. Finally, it is expensive in the measure that if visits to pediatric emergencies continue to rise, the total number of emergencies will reach unattainable heights. If an optimization program is not implemented, it will become necessary to allocate complementary resources, which are becoming a rarity.

3.5.5.2. Overcrowding risks at pediatric emergencies

Despite the efforts deployed, PES show problems similar to those of HES. Besides, the 2014 report from the *Cour des Comptes* of 2014 describes the gravity of the situation for pediatric emergencies, particularly for those children below 1 year of age (Figure 3.5).

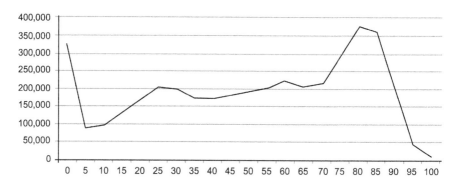

Figure 3.5. *Number of hospitalizations in 2012, resulting from a visit to emergency services, classified by age (5-year leap) (source: [LER 14])*

For some of the patients, the reason for resorting to emergencies is linked to the accelerated gravity of an existing known illness, which could have been solved earlier.

As remarked by Boisguérin and Valdelièvre [BOI 14], the visit rates are particularly important at extremes, especially for the under 1 year (2‰), and (1.5‰) for the over 85 years old, with an average of (0.8‰) for all the population. The reasons for the emergency visits vary highly depending on the patient's age. For children under 15 years old, they are often concerned by pathologies in the area of ORL-respiratory, gastroenterology and rheumatology (Figure 3.6).

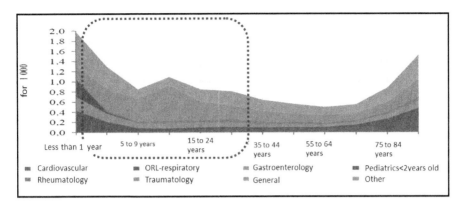

Figure 3.6. *Rate of visits to emergency services, by reason and age (source: [DRE 13]). For a color version of the figure, see www.iste.co.uk/zgaya/logistics.zip*

Beyond general difficulties associated with the hospital system, PES, the same as adult emergencies, present certain features of an organizational nature, particularly:

– Waiting time: waiting time at emergencies is recognized as a pertinent objective and constitutes a relevant quality indicator for these structures. The prolongation of waiting time goes against patients, particularly the most fragile patients. In this way, the overcrowding of emergency services may provoke an excessive length of stay. Research has made it possible to establish the link between waiting time and care quality. According to [COL 07], waiting time constitutes a composite indicator that enables us to account for the capability of a service to adapt to what is happening bottom-up and top-down, as well as other constraints.

– Activity peaks: one of the features of the activity of emergency services (ES) relies on the fact that it often presents activity peaks. These are of two types [DAK 11]: intrinsic to the service or extrinsic, corresponding to the admission of new patients. The first category mainly depends on modifications of the patient's clinical condition. The second category can be explained by the fact that admissions at emergency services are not programmed, and consequently, the patient flow to ES is irregular. This irregularity impedes any kind of a priori planning and elicits difficult problems of organization to the managers of these units.

– Numerous participants: numerous professionals are around with one patient and must exchange an important amount of information in order to coordinate their actions. As care activity cannot be interrupted, even during the night, the continuity of care is guaranteed by the relay of many teams. Working hours are distributed among many shifts.

At ES, we observe a strong dispersion in the average number of patients nursed each day: for 28% of the structures, less than 40 patients per day; for 43%, between 40 and 80 patients per day; for 18%, between 80 and 120 patients per day and for 11%, more than 120 visits per day (Table 3.3).

Establishments		Emergencies		Number of daily visits			
				<40	40-80	80-120	>120
Public	494	General	479	116	179	109	75
		Pediatric	106	43	48	11	4
Private, non-lucrative	41	General	41	12	23	6	
		Pediatric	4	1	3		
Private, lucrative	120	General	119	37	72	8	2
		Pediatric	1				1

Table 3.3. *Enumeration of hospital emergency structures according to the center status, the type of authorization and the number of daily services*

All these elements invite a reflection about the management modes of patient flow at the PES throughout the circuit, from the moment the patient is nursed until he exits the service. For all these reasons, and due to their complexity, emergency services have been the object of a number of works dealing with the problems associated with modeling, simulation and optimization of patient flow.

3.6. Hospital and healthcare information systems

Hospital and healthcare information systems are considered a key piece in the care system [FAB 04]. One of the current challenges is to allow, by using efficient and inter-operational information systems, the exchange of information between all actors, in order to improve the follow-up, evaluation and control of activities during the care process. It is essential to implement networks favoring semantic inter-operability and collaborative work for distributed medical applications, taking into account the specificities of the field (vital reliable information, specificity of certain actors, existence of itinerant activities, multiple sources of information, information used by many actors, etc.).

These systems may concern the circulation of a medicine, medico-economic control (T2A, common classification of medical deeds, indicator follow-up, etc.), external (telemedicine, city-hospital networks, imagery, HAD, etc.) and internal (access to patient's file, access to information and images, etc.) communications. They raise numerous types of problems: conception and implementation, use, appropriation, user application acceptance and IHM.

3.7. Analogy between conventional and healthcare logistics

As its main mission is to provide the best possible care to the patient, the care system is described as a true production system with a long list of professions. These numerous care professions possess a deontology, a professional culture and a specific hierarchical organization. Besides the typical hospital professions, we can add a broad range of professions indirectly associated with care activity (administrative, technical, informatics, etc.). Apart from this diversity, the complexity of care activities is significant and numerous: multiplicity of missions, hierarchy, compartmentalization, heterogeneity of equipment and facilities. A care system is like a supply chain (SC), which connects many establishments, actors, resources, etc. This system is described in section 1.3.5.3.

Despite the fact that the concept of supply chain management stems from the industrial field, the health field needs to exploit performances and scientific and technological advances used in the industry. This exploitation is not trivial because of the human factor involved in the field of health. The main difference between conventional and health logistics is that the latter pivots around a central human actor: the patient. This is at the center of the whole care system.

Table 3.4 introduces an analogy between conventional and health logistics, by applying the previously introduced basic concepts.

Logistics concept	Conventional logistics	Health logistics
Logistics (section 1.3.1)	The means making it possible to have the right product, at the right place, at the right moment, at the lowest cost and with the best possible quality.	The means making it possible to have the right product (patient file, medicine, staff, etc.), at the right place (operating theatre, reanimation room, central drugstore, etc.) at the right moment, at the lowest cost and the best possible patient care.
Supply chain (section 1.3.5)	The system thanks to which organizations provide their products and services to the final consumer.	The system thanks to which healthcare establishments and medico-social centers provide their care products and services to the patient.
Logistics process (section 1.3.2)	A set of operations and events using all available means (human, material and information resources) mastering the corresponding flow (section 1.3.4) These means can be human (staff), material (facilities and equipment) or methodological (techniques and methods) and must be efficiently coordinated in order to attain the best possible service level.	
	The aim is to provide a set of services at the lowest cost and the best possible quality to the beneficiary (final customer or intermediary actor of the chain).	The aim is to perform a set of caring services (direct care tasks) or services associated with care (e.g. preparation and sale of medical products) at the lowest cost and with the best possible quality for the patient and the care actor (doctor, pharmacy, nurse, etc.).
Supply chain management (section 1.3.6)	Optimization of SC management. This consists of managing all logistic processes of the extended enterprise, from the supplier's providers to the client's customers.	Optimization of SC management. This consists of managing all logistic processes of the health system, from the health product suppliers (medical team → care, healthcare products enterprise → medical material, pharmacy → medicines, etc.) to the patients.
Flows (section 1.3.4)	Three categories: physical (material and human resources), information and financial.	*Physical (material and human resources)*: Medico-surgical material (X-Ray equipment, syringes, bandages, etc.); Medicines and diverse molecules; Testers and samples; Hotel material (food trays, beds, sheets, diverse goods, etc.); Cleaning material; Person flow (patients, staff, visitors). *Information*: All transferred data for managing the indicated flows. *Financial*: Every financial transaction associated with every type of flow.

Table 3.4. *Analogy between conventional and health logistics*

3.8. Conclusion

In this chapter, we have introduced the healthcare sector at a regional, national and international level, illustrating health organizational problems through a literature review. For the solution of these problems, using an analogy between conventional and health logistics, we have shown that the logistic tools and methods already deployed in the industrial sector can be adapted to the field of health. In the following chapters, we will introduce the research works conducted by the OSL/CRISTAL team, using advanced methods for implementing decision support systems to better manage health logistics flows.

In this context, using the A3C-2SL architecture introduced previously (section 1.9) for the resolution of logistics systems is also crucial for the health field. This architecture needs deeper research work and development at the level of the three layers. Indeed, apart from the mathematical formulation work of constraints and health criteria, as well as the optimization algorithms specific for the health field (layer 1), modeling based on communicative agents (layer 2) establishes collaborative optimization for scheduling and dynamically orchestrating patients' pathways. This procedure is represented by a collaborative Workflow (layer 3).

4

Collaborative Workflow for Patient Pathway Modeling at Pediatric Emergency Services

4.1. Introduction

In this chapter we present collaborative workflow through some definitions that help us to characterize and classify the different categories. We will examine the functioning modes of Pediatric Emergency Services (PES) of the University Hospital of Lille (CHRU). Workflow for patient pathway modeling will be established at the PES. It is important to study the constraints that a PES context imposes to Workflow management models. Finally, in the last section, we will introduce the interest of multi-agent modeling for the scheduling and orchestration of the suggested Workflow. As MAS are interesting for workflow management, we will also introduce a detailed study of them in Chapter 5.

4.2. Definition of workflow

Many definitions of the workflow approach have been proposed by the specialized literature. Among these definitions, we can quote a definition introduced in Hales [HAL 91]: "Workflow management is done by a proactive system for the management of a series of tasks which are transmitted to the appropriate participants in the right order and which are completed under allocated time".

Chapter written by Sarah Ben OTHMAN, Inès AJMI and Alain QUILLIOT.

This definition highlights the main terms:

– a proactive system which generates objectives and satisfies them, and whose behavior is conditioned by the system's internal considerations, as well as by exterior events;

– adequate participants are called by the actors or the resources. These participants can be software entities or persons;

– the task sequence which is generally associated with previous constraints.

In order to generate a workflow, it is necessary to go through two stages: process generation and process control and implementation. Figure 4.1 describes the relationship between these two stages.

After a functional analysis phase, preliminary conception is established for a workflow. There are many methods and graphic tools for conceiving a workflow diagram, as well as for verifying the validity of the model [MÜH 02].

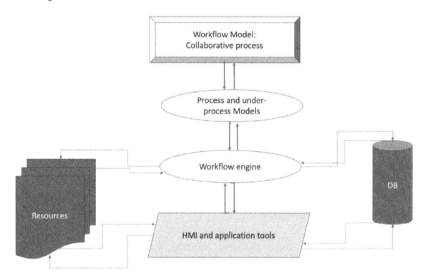

Figure 4.1. *Workflow management phases*

The execution phase is based on the creation of workflow diagram instances, executed using the appropriate data and adequate resources [STO 03]. A set of variables and parameters are necessary for the execution of each instance. This set makes it possible to determine which tasks have already been performed and with what results [KAM 98]. The workflow engine allows for carrying out the execution phase. This consists of using software applications, HMI and available resources for coordinating the execution of processes and activities. Processes are conducted from one informatics area to another, and each time a process phase has been completed [ZGA 10].

4.3. Why use a "workflow approach" in health?

In recent years, health establishments have entered the era of digitized collaboration, and to remain competitive, they are obliged to permanently improve the quality of their services (Chapter 3). The accessibility of an increasing volume of data and the integration of varied software create new demands regarding collaborative management tools, be it inside the hospital or in the context of inter-hospital cooperation.

Health workflow is an approach to model and manage medical or administrative activities at a health center, involving many actors, documents and tasks. It consists of working models that help coordinate the activities of each medical staff member and to ensure their perfect interconnection, relying on computer systems and existing databases.

Beyond the benefits that can be obtained from the implementation of a Workflow health management system, the restructuring of hospitals in view of their adaptation to digitization can improve a number of key performance features:

– efficiency: hospital patient pathway modeling eliminates useless medical tasks;

– control: a better management of medical processes is obtained thanks to the use of standardized collaboration tools and thanks to the availability of tracks for auditing;

– better patient pathway: builds coherent processes, providing gain in terms of quality service;

– work process improvement: a "flow"-oriented vision provides process simplification compared with role-centered paradigms.

A health workflow approach has the following advantages:

– flexibility: workflow control of medical processes enables their dynamic modification, when medical protocols change;

– optimization: optimal solutions, for example, task organization, can be mathematically and algorithmically calculated, which helps improve performances when compared with informal solutions;

– security: access rights are strictly defined beforehand.

Thanks to these advantages, using workflow for managing medical processes at health centers is largely accepted today. The evolution of health establishments is clearly oriented toward a higher level of digitized integration, regarding interorganizational processes (such as sub-contracting, hospital supply chain management or inter-hospital cooperation).

4.4. Description of a workflow diagram type

The activities of a classical workflow diagram are linked by transitions between many different processes of a given instance. There exist different types of conditions and routes associated with these transitions:

– junction: many converging control threads;

– iteration: one same process sequence is repeated;

– Boolean precondition: the departure point for a set of processes;

– Boolean post-condition: the criterion that helps stop a process sequence;

– parallel routing: many processes being executed in parallel;

– scheduled routing: processes are executed in a sequential manner;

– multiple connection: partitioning one control thread in many parallel threads;

– meeting point: meeting point for many control threads;

– conditional connection: guiding many activities to execute one control thread;

– transition: passing from one process to another after verifying certain conditions.

4.5. Health collaborative workflow

This type of workflow is perfectly suited for health modeling. In collaborative workflow, optimizing the execution of a Workflow instance is less important than the capacity to react with flexibility to unforeseen modifications; medical processes are defined in an informal manner. Medical staff may intervene at any moment for decisions concerning the routing of processes.

A health collaborative workflow approach must enable medical staff or teams to cooperate in order to achieve the common objective of patient care.

Execution of a process in the case of collaborative workflow is divided into five stages:

– request (is there any available human or material resource?);

– negotiation (which resource can be liberated to execute urgent tasks?);

– agreement (yes, the resource is available);

– allocation (the resource is employed for executing one task in the process);

– verification (has the task been correctly executed?).

4.6. Inter-operability concepts for health collaborative workflows

The control and management of a collaborative workflow involving many actors can be distributed in different ways:

– Capability sharing: medical tasks are allocated to the medical staff according to predefined rules, through a global medical protocol. Task execution can be distributed among different teams.

– Chain execution: workflow is divided into many parts which are sequentially executed by different medical teams. As soon as a team has finished its part, it passes on the control phase to the following team. Contrary to capability sharing, execution and control phases are distributed.

– Hierarchical execution: workflow management is distributed in a hierarchical way among actors. For example, a doctor passes on the control of one section of his workflow to a nurse. The nurse executes the allocated tasks and passes on the control to the doctor after having finished execution.

– Workflow execution transfer: each actor possesses a copy of the complete workflow diagram. Only workflow instances are transferred among actors, in order to balance responsibility sharing because a task must be performed by medical staff who have the required skills (Figure 4.2).

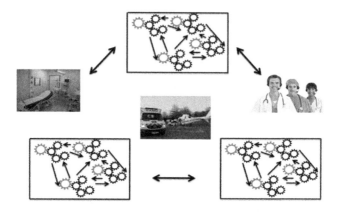

Figure 4.2. *Workflow transfer instances*

– Distributed execution: the parts of a workflow diagram are distributed between medical actors and can be active in parallel. Actors do not always know the diagram and the condition of Workflow instances of their partners. Communication between instances is done under a common protocol (Figure 4.3).

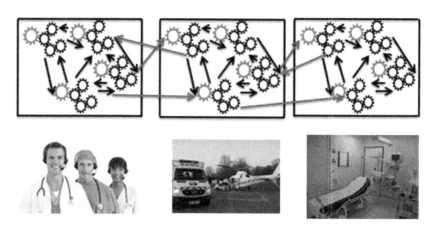

Figure 4.3. *Workflow distributed execution*

Collaborative workflows are easily inter-operable in health and are thus best suited for modeling patient pathways in PES, be it in normal or in overcrowding functioning mode (Chapter 3).

4.7. Patient pathway description for PES (CHRU[1] de Lille)

The patient pathway at a PES starts in one of two ways:

– *Emergencies*: the mode of entry for the majority of patients, whether they have come by their own means (parents, family, etc.) or in an ambulance which may not have informed the hospital of its arrival. This entry is common to all individuals entering the CHRU through *Emergencies*, children and adults. Parents or other family members must follow the administrative admission procedure before being redirected to PES.

– *SAMU*[2]: the arrival of a patient directly via the ambulance entry platform, after which the patient is immediately redirected to PES, while family members are sent to the administrative office in order to complete hospital admission files.

If the patient has entered the hospital via *Emergencies*, the moment they arrive at PES, they undergoe a first diagnosis established by a nurse, which determines priority order to access services. This priority order depends on a 5-level scale for immediately treating the most severe cases. These emergency levels, detailed in Chapter 3, are represented by the CCMU (*Classification Clinique des Malades aux Urgences*).

After a more or less significant waiting time, depending on service saturation and emergency condition, the patient is finally treated by a doctor. The complete process is summarized in Figure 4.4.

After the medical diagnosis has been established by a doctor, he or she may choose – depending on the medical condition and available infrastructure – to let the patient go back home, to keep them in observation, or even to hospitalize them. During observation, the patient is required to stay in the hospital for a short time because of the lack of top-down structures allowing for reception. Generally, there are no available beds under ongoing supervision at a hospital. What is more, there is no central

---

1 University Hospital of Lille.
2 Ambulance Emergency Services.

patient alarm. Equally, if the doctor decides to hospitalize the patient for a period longer than 12 hours (maximum short hospitalization stay), he or she is obliged to keep the patient in the PES waiting for a bottom-up structure to be available. Finally, patient reorientation to less busy departments or hospitals, if any are available, is very rare, due to the lack of communication between hospitals, and the difficulty in knowing the degree of overcrowding among health institutions. This process is clearly summarized in Figure 4.5.

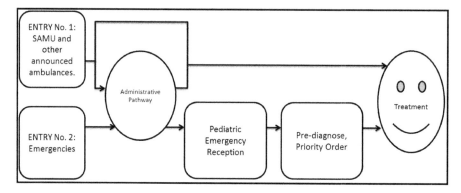

Figure 4.4. *Patient pathway*

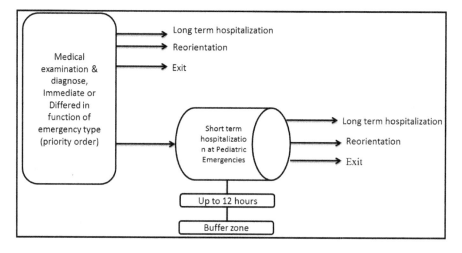

Figure 4.5. *Diagram of possible patient orientation after initial medical diagnosis*

The flow of patients entering the service is totally random. However, the average amounts are sufficiently cyclic and redundant to generate an entry flow model, at different time scales: according to the season, moment in the day, certain particular events, etc.

4.8. PES infrastructure

4.8.1. *Evaluation*

The under study PES covers a 450 m² surface at a health center. While this may seem small, it is important to note the existence of two other equivalent structures in the city and that the patient intake at the CHRU per year is modest. Despite this, PES is becoming insufficient in terms of space, which also leads to the need for polyvalent staff.

The PES includes:

– a reception;

– 10 cubicles, with simple or double beds;

– one plaster room and one suturing room (also available as cubicles);

– one emergency room, that is, a room for monitoring patients in need of constant observation (must be liberated as soon as possible);

– one waiting room (in the corridor).

This layout is graphically represented in Figure 4.6.

PES can equally rely on top-down structures inside the hospital, which also play a part in the pediatric service. PES has 170 top-down available beds. In addition, the hospital is equipped with three operating rooms:

– a "traumatology" room;

– a "heart surgery" room;

– a "neurosurgery" room.

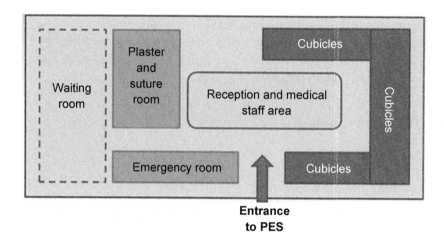

Figure 4.6. *Physical architecture of PES*

4.8.2. *PES management*

Unfortunately, there is currently no reliable system allowing for an efficient management of available spaces. There is also the problem of "falsely free" beds, that is, beds which have been booked, or which are unavailable due to the presence of a contagious patient, or the fact that the room has not yet been cleaned after a patient has been discharged.

4.8.3. *Staff*

The medical team at PES consists of:

– two care pediatricians (specialist doctors);

– one generalist intern;

– two pediatric interns;

– one surgery intern;

– two to three nurses, two in the morning, three in the afternoon, two at night (one of them at the reception);

– two childcare assistants.

This is enough staff 90% of the time, that is, when the service is under normal activity. However, it is not sufficient during busy periods. This problem cannot be solved by increasing the workforce during intense working periods for juridical and trade union reasons. In fact, nurses working at hospitals cannot do temporary work and must regularly work every night between 6 pm and midnight (one of the most difficult periods).

4.9. Collaborative workflow for modeling patient pathway in the steady state

"Workflow" is the representation of a sequence of tasks or operations performed by an individual, a group of individuals or an organization within a time frame. These tasks are articulated around a predefined procedure and generally have a common objective. More precisely, the term "flow" refers to the passing of a product, document or information from one stage to another. We distinguish many aspects in the concept of collaborative workflow. In fact, this approach can represent interactions between tasks in the form of information exchange between different actors, providing each with the necessary information for task performance. It describes the circuits and the validation modes, as well as the deadlines to meet for each task in each process. What is more, the notion of "workflow" is based on BPMN[3], presenting a standardized graphical notation system for modeling business processes. The objective of BPMN is to provide a model that is suitable for all service users.

In order to validate the modeling provided by "workflow", simulation and result analysis are extremely important. Patient pathway modeling represents a pertinent way for managing material and patient flows. This guarantees quick and qualitative care, planning material and human resources for hospitals. This stage formalizes the precise functioning of PES, more easily automating certain flows and identifying optimization zones. Process modeling offers the possibility of creating graphical representations, facilitating process understanding, identifying involved actors and their roles, and ensuring information sharing.

4.9.1. General functionalities of PES

The model presents the organization of a PES in its day to day functioning. PES is in charge of children and adolescents with medical and

3 Business Process Model and Notation.

surgical emergencies. According to their condition and pathologies, they are directed to the appropriate care unit to be treated. Studies and consultations carried out with the medical experts have led us to define individual pathways, associated with a type of pathology and certain conditions.

While considering these small variations, we can observe that typical pathways are representative of this service mode functioning. There are five big care categories existing at the core of the structure:

– cubicle consultations: pediatricians are briefed in the context of specific monitoring (chronic diseases, post-hospitalization, premature babies, etc.) or unscheduled consultations (according to the specialties indicated by doctors);

– short-term hospitalization: enables children to receive all surgical emergencies under the joint responsibility of a pediatrician and the emergency doctor;

– external care: includes suturing and plasters;

– traditional hospitalization: cares for children suffering from acute or chronic diseases, localized injury, intoxication, psychological difficulties, disabilities etc. Led by a specialized team and in a context particularly adapted for children, in pertinent reception conditions and with the proper care.

– emergencies: operations carried out in an operating room.

4.9.2. Process modeling

In order to build a complete diagram, conforming with reality, it is essential to represent the patient pathway from the moment the patient arrives until the end of his or her treatment. However, in order to obtain an accurate representation of the waiting time, and to later improve it, it is necessary to have a larger and more global vision of the patient pathway, which contains bottom-up processes in the admission of emergencies. These processes have a knock-on effect and prolong the waiting time. Owing to this, we are obliged to take all administrative procedures into consideration, such as patient registration and the multiple orientations, that can take place and which directly influence the pathway.

4.9.3. *Global process*

Figure 4.7 represents PES modeling under the steady state. It is composed of four processes:

– one main process that globally describes patient pathway in the service;

– three sub-processes detailing the functioning in each unit: short-term hospitalization (UHCD[4]), external care and emergencies.

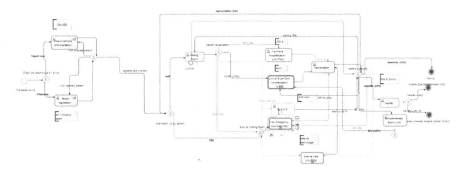

Figure 4.7. *PES process modeling. For a color version of the figure, see www.iste.co.uk/zgaya/logistics.zip*

4.9.4. *Sub-processes*

The process of describing the patient pathway, from the moment one enters the hospital until they leave, gives an image of the functioning of the service. To be more precise, this has been completed by sub-processes representing care in UHCD (Figure 4.8), in external care (Figure 4.9) and in emergencies (Figure 4.10). This makes it possible to detail a task without complicating the general diagram. For example, UHCD entail a higher number of examinations and verifications than a simple consultation. The transitions related to this type of care could overcharge the diagram if they appeared in the final process, making interpretation difficult (it would be impossible to understand the diagram and the UHCD "task"). We have thus created a specific sub-process: when a patient arrives at this stage, he or she is directed to the entrance of the sub-process, and when he or she reaches the end, he or she is reinjected into the main process.

---

4 *Unite d'Hospitalization de Courte Durée* (short-term hospitalization).

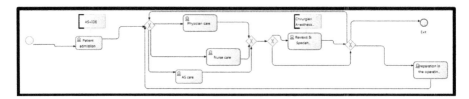

Figure 4.8. *PES process modeling (UHCD sub-process)*

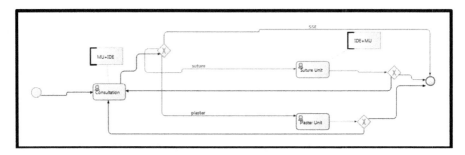

Figure 4.9. *PES process modeling (external care sub-process)*

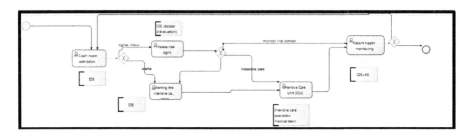

Figure 4.10. *PES process modeling (emergencies sub-process)*

This type of modeling considers all types of care and orientations at the core of the structure, as well as registration tasks and form filling. These forms evolve at each stage of patient pathway. They constitute the patient's file. This file ensures the traceability of all the tasks performed. It equally facilitates passing information and communication between care actors and patients. In addition, this model takes into consideration the notion of waiting time, which is modeled by concrete tasks such as "Wait for Orientation", which represents the amount of time spent in the waiting room in order to obtain an accurate image of the reality.

4.9.5. *Modeling under the overcrowding state*

During peaks in activity, PES services become saturated. The waiting room is no longer big enough, families cram into the corridors with their children and waiting times increase. In fact, it may take longer than 5 hours during periods of high affluence, while an average waiting time is less than 2 hours. The service moves into a new functioning phase, which we call "overcrowding" phase, when the aim is of alleviating patient flow.

A flexible and practical mindset should be implemented to run the service, to minimize the inconvenience of wasting available resources. During peaks, doctors take the initiative of using any and all available resources, no matter what their theoretical specificities are, in order to respond to patients' needs. Resources such as rooms or beds, and even certain human resources, are generalized. For example, a consultation normally carried out in a cubicle is then moved to the suture room.

4.9.6. *Collaborative workflow for patient pathway during overcrowding*

The state of PES during overcrowding has led us to define a new patient pathway, which considers the possibility of using resources that have not been assigned to tasks when doctors can relocate them (Figure 4.11). These resources mainly concern service rooms: we focus on this aspect for redefining patient pathway. In this way, the entry – which has so far remained administrative – does not change at all with regard to permanent modeling. In fact, it is important to correctly register the patient according to administrative and legislative criteria. With regard to the exit, this cannot be modified because there is no reason for reoriented patients to come back to the service. On the other hand, exits are, as entries, purely formal administrative stages which imply purely external bureaucratic services and which serve to ensure correct patient monitoring, from admission to exit. These services enable the hospital to be prepared in case of conflict with a patient or following complications in the patient's condition.

Each service (simple consultation, operation, plaster, short-term hospitalization, etc.) can be performed in different rooms. It is thus convenient to have classified patients depending on the care they have to be provided with and to separate them again depending on the room where they will receive treatment. This operation multiplies the possibilities of care and makes the pathway more complex. On the other hand, classification must be correctly modeled, because it does not follow explicit rules. Indeed, doctors choose, at the last minute, which room the patient shall be led to. For example, emergency rooms are only used in the last resort, because they are devoted to treating patients whose life is in danger. Evidently, the preferred room will be the one indicated for a special type of care. Then, other rooms are automatically assigned, without particular priority.

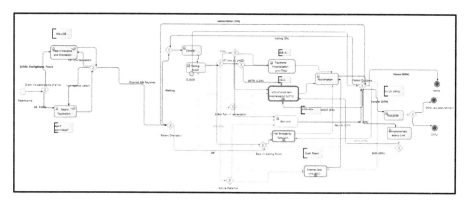

Figure 4.11. *PES process modeling under the overcrowding state.*
For a color version of the figure, see www.iste.co.uk/zgaya/logistics.zip

4.9.7. *Choice of transitions and configuration*

There are many ways of defining how to move from one task to the next, or the direction to take after a node (junction point where flows can later be led to different tasks). The tool used gives the possibility of defining a direction in two ways: conditional or probabilistic. Their impact on the process and the modeled pathway being different, we mix two types of transitions in our model.

The conditional method consists of defining variables and distributing flows, arriving at the nodes according to the value of these variables, which correspond to the arrival path. For example, it is useful to ensure that a stage has been successfully completed (for instance, the case of a patient in a non-critical condition who may have forgotten to register at the administrative office and who would be reoriented in that direction in order to respect admission protocol), or that certain situations be correctly monitored because of the tasks they imperatively imply (a patient arriving at emergencies will be automatically redirected to an operating room). However, for those patients who are not labeled as emergencies, there are no previous conditions indicating what type of care they will receive. In order to correctly distribute patient flow across different tasks, we use the probabilistic method, attributing a probability value to each task, susceptible of being used by a patient. The percentages we attribute to these transitions derive from the exploitation of databases provided by the hospital, which sum up a year's patients' entry and exit times, as well as the reason for their coming to emergencies.

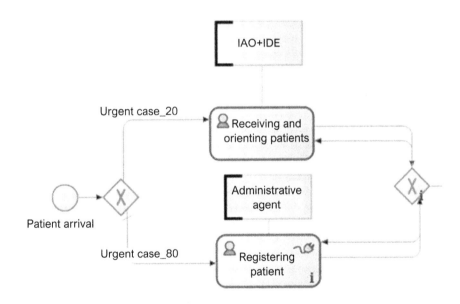

Figure 4.12. *Example of intersections or nodes*

In this section, introduced by Figure 4.12, we use the "eXclusive OR" connection to register patients. This means that from the patient's arrival, he or she is directed to registration, either from reception or by orientation, according to the seriousness of their case. However, the patient will not be able to move on to the next stage unless both tasks ("registration" and "orientation") have been validated. For this, we have used the probabilistic method, which serves for correctly distributing flows. Chosen percentages result from the study of databases provided by hospitals (e.g. urgent case_20 = 20% in Figure 4.12 corresponds to the probability of patients arriving with an urgent case).

4.9.8. Connectors

The chosen tool possesses a vast choice of connectors for information systems, databases and different content, even for social networks. In order to implement an application that corresponds to our needs, we make use of built-in connectors.

In our application, we have used two types of connectors:

– the first is based on data that make it possible to link the Workflow model to a base that we have created with the data base management system MySQL. We have used this connector for registering or automatically upgrading patients' information, from the moment the patient fills the registration form;

– the second is linked to emails. It is useful for sending an email with a summary of the patient's file according to the information he provided during registration. This email is sent when the patient is formally discharged.

4.10. Agent-oriented approach for collaborative workflow

MAS are provided with autonomy and reactivity. Under the term "*MAS for workflow*", we gather every distributed approach for executing a workflow that involves cooperative entities. Health collaborative workflow models are equipped with workflow execution features in a distributed manner. In fact, MAS are composed of distributed entities (many services at PES), capable of communicating, which may influence the development of process execution. These entities have a specific condition which is not perceived by other entities. Interactions between different entities may generate new solutions for optimization problems, such as bringing a medical process to a satisfactory conclusion.

Workflow agents can be classified into three categories:

– agents who cooperate at the service of patients. Here, each agent plays a similar role to a physical agent (doctor) in a PES, to improve patient care;

– distributed agents responsible for the reactive coordination of care tasks. They are based on activities, not on roles. Their coordination is managed by a Workflow diagram, without a central execution engine;

– agents going from one "service point" to another. For example, a doctor agent may migrate from a care team to another team, in favor of the patient having priority.

4.11. Agent coalition for executing collaborative workflow

Collaborative workflow helps to model the patient's position and his or her movements at the PES in an accurate way. Owing to the unforeseeable nature of care activity and the uncertain environment in a PES (which requires a dynamic management of information), forming agent coalitions is necessary to solve special problems such as care task planning, resource management and workflow supervision.

A coalition is based on agents playing the role of cooperative actors. Each coalition agent is capable of controlling the execution of a Workflow instance. An instance is composed of constraints and conditions operators (sequence, parallelism, etc.) in order to create complex care tasks. These care tasks can be intertwined at all levels [BEN 15c].

The agents of a coalition must reach agreements in order to decide whether they need services provided by agents from another coalition to improve the health of one or more patients. These agreements are called "negotiation". A negotiation protocol has been proposed, allowing agents to provide the appropriate decisions and to establish their execution conditions (unique or repeated execution of a workflow instance, medical staff allocation, etc.). Negotiation strategies explicitly use two knowledge databases: the first database is declarative and serves for describing objectives and the context of negotiation (e.g. to adapt the care room according to the patient's pathology). The second database includes negotiation rules, for example, allocating the best members in the medical teams to maximize quality of care [BEN 15d].

The objective is to define a negotiation protocol, so that coalition agents can achieve the distributed execution of a workflow instance. We suggest a two-step approach for the management of a collaborative workflow. Agent coalition and the split will enable us to differentiate two distinct phases in the model. The first phase corresponds to the allocation and scheduling of care tasks, which will later serve the Scheduling Agent (SA) for control. The second phase consists of dynamic orchestration, relying on a negotiation process between coalition agents (e.g. SAs and medical staff). The underlying concept is to separate coalitions depending on the nature (predictability) of the knowledge they manipulate.

In scheduling and task allocation phases, the SA analyzes the description of a workflow diagram. This diagram defines the inherent properties respecting care protocols, which are fixed before execution. This includes specifications about existing resources, the list of tasks to be executed as well as the precedence links between these tasks.

The dynamic orchestration phase relies on a dynamic scheduling methodology. Apart from time constraints, the real availability of resources is equally taken into consideration. During this phase, tasks are scheduled according to calculated priority, as workflow is progressively executed.

We will explain these two phases in detail in the next chapter.

4.12. Negotiation protocol between agents controlling a workflow instance

The SA is the initiator of the negotiation protocol shown in Figure 4.13. He or she ignores the information regarding resources to be allocated to care tasks, but possesses all the information about patients, their pathologies and their needs.

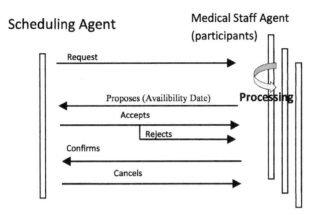

Figure 4.13. *Negotiation protocol between agents*

The SA sends a request to the medical staff agent, considered the best qualified for executing a care task. The message has the following format:

<SA, medical staff, requests, availability date of medical staff, care tasks to be executed at a t moment>

with:

– availability date: corresponds to the earliest date when medical staff is available for executing care tasks;

– care tasks to be executed at the moment *t*: these tasks can be performed by several members of the medical staff. Each member analyzes the request, calculates his real-time availability and sends a reply to the SA, who chooses the most adequate medical staff (he accepts or refuses proposals).

The SA remembers all the information sent during a negotiation cycle. In this way, he or she can know which members from the medical staff must be informed after an execution modification of a workflow instance.

Each SA also negotiates with medical staff members from his or her coalition, to book schedules for executing care tasks. This booking concerns all available schedules during a time period, starting with the time frame available for the workflow at hand. Each booking must include patients and priorities of care tasks calculated previously by the SA. In exchange, every SA coalition member receives a list of occupation rates for each of the required time schedules. At the end of a negotiation cycle, communicated via a SA message, the medical staff mobile members chose the first care task in the planning for executing it at the proper time schedule. Then, they inform the responsible SA, who can later access the resource during the execution of this task. Agents from other coalitions are equally informed about the non-satisfaction of their booking demands for already allocated time schedules.

Once the execution of the care task is over, the SA controls workflow for execution of the next task in schedule and requests the necessary resources he needs. Two situations can lead the SA not to move forward in the execution of his scheduled tasks; either he has not been chosen for accessing a resource, or he must wait for the ending date of the tasks that precede his own. In this case, the SA must wait for the liberation of resources and the end of his predecessor's tasks. A SA having no more tasks to perform is eliminated from the system.

4.13. Global coherence and periodic behavior of collaborative workflow

The execution of collaborative workflow is confronted to two main difficulties:

– global coherence: care tasks to be executed at different PES for the same patient are connected by constraints. The decision to execute a care task is made according to the development of the patient's health condition. To ensure the respect of constraints, it is necessary that the SA communicates the patient's condition to the different places where he is cared for. In order for a task to be executed, it is necessary to verify the availability of human and material resources at the right place and at the right time;

– periodic repetitive behavior: throughout information exchange between agents at the interior of a coalition or between agents belonging to different coalitions, the status of each agent is updated according to gathered information. This update has an impact on the rest of the negotiation process, provoking the status change of an agent. In this way, negotiation processes may oscillate from one state to another. In the following paragraphs, we will analyze these two problems and suggest a solution for them.

4.13.1. Respect of global coherence

One of the problems associated with the nature of collaborative workflow is the untimed propagation of information concerning the care tasks that have already been performed over a patient, and those yet to be done, as a function of precedence constraints. Let us suppose that negotiation deadlines between the entities of a MAS are minimum regarding an available time frame. In that case, precedence constraints are not violated, because the information relative to previous care tasks will be transmitted before the execution of subsequent care tasks in all cases.

It is always very difficult to detect the ending of a negotiation period, even if it is short. The chosen solution is often heuristic: the SA makes some patients wait during a limited amount of time after sending the last message for booking medical staff resources, or the propagation of pertinent information, before triggering the execution process of a care task. Thus, certain messages in course of transfer or treatment can be considered.

4.13.2. Treatment of oscillating states

For each coalition, the SA books resources according to the total estimated duration of the execution of previous care tasks. This estimation is based on the execution probabilities of previous tasks, which may also be needing the same resource. The workflow management system can be in an oscillating state.

For example, we may suppose that a SA allocates a "medical staff" resource for executing a care task whose execution probability overflows the predefined booking threshold, which provokes a decrease in the availability of this resource. The SA responsible for the allocation of the resource consequently warns the other agents who have already booked this new availability. These agents then recalculate their probabilities and encourage subsequent agents to adapt their calculations as well. If the SA belongs to subsequent agents, the probability of executing his or her task may be below the booking threshold, and entails cancelling. Following this waterfall recalculation, "medical staff" agent then informs the SA about the rise of his availability and the cycle begins again.

In order to avoid oscillating states, we introduce an amortization variable at the workflow instance, which makes it possible to progressively reduce the impact of new bookings on the estimation of "medical staff" resource availabilities, in order to achieve a stable situation. Evidently, this phenomenon does not guarantee that the stable situation will be optimal for our multi-criteria objective.

In the following chapter, we will briefly introduce an architecture agent that makes it possible to use these approaches for scheduling and orchestrating the workflow.

4.14. Treatment of generated collaborative workflow decision points

The diagram of generated collaborative workflow has many decision points, in which the execution of ongoing tasks depends on the execution of previous tasks. Decision points are generally graphically modeled by lozenges (gateways).

In our model, validation tasks can either lead to a continuity of task execution, or a comeback to the previous task (e.g. after consulting a doctor, the patient may go back to the waiting room). During the initial allocation of tasks to the SA, we can quote three types of treatment:

– ending points of a Workflow can be considered as decision points. The execution of an instance finishes when a new instance starts, depending on the decision made by the SA;

– decision points can be classified and they propose tasks to the SA according to the possible results of these decisions. A Workflow instance is generated as a function of credible task sequences. For example, if a patient's pathology is positively validated in 90% of the cases, the SA only orchestrates the "positive validation" workflow branch. Any disturbance is considered in the execution of a workflow instance is seen as a less credible sequence. If the decision is made, the SA may generate an exceptional treatment in the current step. This decision is characterized by the SA recuperating control at the previous step;

– when workflow execution begins, all instances of all the branches of all decision processes are parallel. If activation conditions are not verified, the corresponding SAs remain in the system without executing their tasks. The inconvenience of this approach is overburdening the system, due to resource consumption of inactive agents.

The choice of a solution, explained in Chapter 5, strongly depends on the specific context of workflow. Deeper analysis is necessary in order to evaluate the advantages and drawbacks of each approach, for a given application.

4.15. Summary

Collaborative workflow, which we have adopted for modeling PES patient pathways, is based on the cooperation of software entities equipped with an acting capacity, that is why we call them software agents. The most recent approaches are appealing to the concept of mobile agent, which allow information transfer for executing workflow instances. Communication between agents, the capacity to react to events, as well as the implementation of communication protocols constitutes the main features for workflow orchestration. In order to be capable of adapting to unforeseen situations, a workflow management software must integrate a scheduling module. Despite the fact that inter-service cooperation demands a completely distributed execution engine, a small number of works study the problem of dynamic and distributed scheduling, particularly in the presence of the constraint which demands respecting the skill of each medical actor.

4.16. Agent activities for collaborative workflow

Owing to their different properties (autonomy, communication, interaction, cooperation, etc.), all the actors of the multi-agent architecture suggested for modeling PES structure can have different activities. In fact, at a given moment *t,* agents can negotiate among themselves, and others can transmit instructions while others are executing tasks.

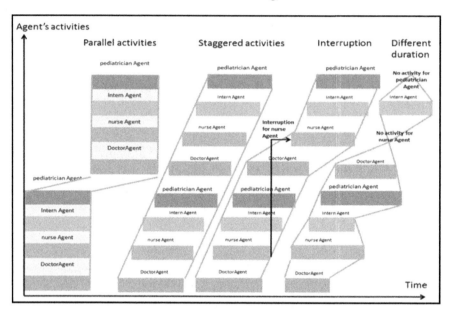

Figure 4.14. *Collaborative agent activities. For a color version of the figure, see www.iste.co.uk/zgaya/logistics.zip*

In Figure 4.14, we identify four types of activities:

– Built-up activities: All the agents work at different paces. An agent is able to start a new activity that can overlap with an activity in progress.

– Parallel activities: In the example, the four agents (medical staff members) work simultaneously. Each agent must wait until an activity finishes, before starting a new activity.

– Split-up activities: An agent may interrupt an already begun activity, if he receives an update or a more urgent care task to carry out.

– Different-term activities: The activities of agents may have different lengths, which depend on the category of the care task in progress.

In order to determine the number of care tasks managed by an agent, we have fixed two objectives:

– To minimize the temporary interdependence between care tasks managed by different agents, in order to reduce the number of exchanged messages and thus respect precedence constraints. There is a particular case, if an agent manages only one task, all the precedence constraints become "inter-agent" constraints, implying an important number of messages.

– To minimize the global execution time for workflow instances, paralleling task execution and thus optimizing the use of "medical staff" resources. A "medical staff" agent cannot be in two places at the same time, for spatial–temporal coherence reasons. It is then forbidden for an agent to simultaneously execute two tasks. When we create only one agent for managing all care tasks, the global execution time of the Workflow is equal to the sum of the duration of all tasks.

As a consequence, an optimal use of "medical staff" resources and a minimization of the global execution time need to allocate different agents to different tasks, which are executed in parallel, via the use of different "medical staff" resources. In the same way, a SA forbids the execution of two parallel care tasks, using the same "medical staff" resource. This is a pertinent case, despite the hypothesis of the indivisibility of resources. The mathematical formulation of these constraints will be detailed in the following chapter.

4.17. Conclusion

The use of workflow tools for managing processes in health establishments is largely accepted today. However, its use in the inter-hospital context bumps into numerous obstacles, for example, the difficulty in transferring a patient file between the different services and the lack of standardization in the interconnection of different local workflows.

Workflow tasks are subject to temporary constraints (precedence constraints, deadlines, etc.) and resource constraints (human or machines). Considering the distributed nature of a collaborative workflow, unforeseen events may happen at every moment. Thus, the coordination of the execution of tasks must be done in a dynamic manner, in order to neutralize or to reduce the impact of disturbances.

In the following chapter, we will study and develop a multi-agent architecture that constitutes a permanent solution for managing the collaboration between PES medical staff in a reactive and dynamic way. A large number of studies in this field use this type of methodology. However, we observe that none of these deal with the problem of MAS scheduling and orchestrating workflows in a dynamic manner, with the additional constraint of having to respect the data confidentiality.

5

Agent-based Architecture for Task Scheduling and Dynamic Orchestration Support

5.1. Introduction

The problems at Pediatric Emergency Services (PES) are primarily related to unpredictable patient flow. Overcrowding at these services provokes considerable waiting times, and all indicators prove that reducing overcrowding is not always easily controlled. To face these problems, we suggest a mathematical formulation of the scheduling problem and an open and dynamic Multi-agent System (MAS), for managing patient care-related problems. In this chapter, we shall focus mainly on the behavior of the Scheduling Agent (SA), which belongs to this architecture. The SA makes it possible to optimize patient flows and to schedule the activity of care processes previously represented by a Workflow. The SA builds optimized schedules for medical staff in order to guarantee a controlled execution of the system, which ensures quick and qualitative care by managing the hospital's resources. In the last section of this chapter, we will study some approaches of Workflow dynamic orchestration done by a SA in order to optimize different criteria used during Workflow execution.

Chapter written by Sarah Ben OTHMAN, Inès AJMI and Alain QUILLIOT.

5.2. Mathematical formulation of the scheduling problem at PES

Collaborative Workflow tasks, described in the previous chapter (Chapter 4), are subjected to temporary constraints (precedence constraints, deadlines, etc.) and resource constraints (medical staff or material). Currently, workflow tools possess a distributed execution engine, but this lacks a scheduling module and does not take competencies into consideration. In the specialized literature, a small number of works deal with the problem of health dynamic and distributed scheduling, particularly in the presence of distributed care constraints and multiple competencies.

The problem of scheduling multiple competence tasks is a highly combinatory optimization problem, analogous to a flexible Job Shop scheduling problem which takes competencies into account. We consider two approaches for its resolution. The first is to sensibly allocate medical staff to tasks. The second is task scheduling, according to their execution and starting dates, considering the different constraints and optimizing a number of criteria. These can be taken into account in the analysis and solution of the scheduling problem. In fact, four criteria are considered: patient global waiting time reduction, balancing the working load of medical staff, minimizing the total charge and minimizing total execution time of care tasks (C_{max}).

The scheduling problem considered must take into account three types of care operations: programmed operations, non-programmed operations and urgent non-programmed operations. In this way, the practitioner must take into account the mobility of patients between care sites (MRI, radiology, biological examinations, etc.). The preemption of non-urgent care tasks in favor of urgent care tasks is authorized. The non-urgent care tasks can be executed in sections for some periods, by different members of the medical staff. Care operations for the same patient are classified into two categories: split-up operations and non-split-up operations. Each member of the medical staff has a set of authorized period treatments and an allotted set of boxes for executing care operations. Periods may overlap and boxes may not be available when needed.

Hard constraints are:

– A medical staff member cannot treat two patients in the same period or in two overlapping periods.

– A medical staff member cannot be attributed to one or many periods.

– An indivisible care operation must be allocated to a unique box.

– Treatment duration of a patient can be equal to the duration of the period in which it is affected or the addition of many consecutive or separated periods.

– The capacity of a box must not be exceeded at any period.

– The addition of the number of patients in the different sites of PES (patients sent to radiology, echography, patients in the box, etc.) must be equal to the total number of registered patients.

– A box can be used at only one period belonging to a set of overlapping periods.

– Each care operation must be allocated only once.

Solving the scheduling problem of multiple competence tasks is not possible unless all hard constraints are respected. The quality of the solution is measured using subtle constraints.

When a light constraint is not satisfied, a penalty is applied. Subtle constraints used for measuring the quality of the solution vary from one PES to another. The following definitions briefly describe the subtle constraints used by practitioners at PES:

– The consecutive allocation of different operation boxes to the same patient must be avoided.

– Two members from the medical staff m_k and m_l, with patients in common, are placed at two consecutive periods during the day.

– Each member of the medical staff must be authorized to leave the patient after the care operation has been performed.

– Under an overcrowding situation, going from one care room to another one requires authorization.

– The preemption of non-vital care operations is authorized. In fact, in case of emergency, an operation attributed to a member of the medical staff can be reallocated to another member, if the latter possesses the necessary competences for its execution.

– The allocation of medical staff members at different sites must be authorized.

– In face of a critical situation, medical examinations can be performed in the corridors of the PES.

Hard constraints must be respected. However, subtle constraints may be violated in tense situations.

In the following sections, we will define a mathematical formulation of different variables, constraints and optimization criteria that will be useful for providing an optimal scheduling of tasks.

5.2.1. Parameters

5.2.1.1. Parameters associated with care operations

NP: an N set of patients to be treated, $NP = \{P1, P2, ..., PN\}$;

MS: an M set of medical staff members, $MS = \{m_1, m_2,.., m_M\}$;

k: medical staff member index m_k;

$O_{i,j}$: ith care operation corresponding to the patient P_j. We represent the patient P_j by the number of his care operations. Thus, for each $P_j \in NP$, $1 \leq j \leq N$, $P_j = \{O_{i,j}, 1 \leq i \leq n_j\}$ n_j number of operations corresponding to the patient P_j;

P_j^s: sub-set of divisible care operations for the patient P_j, with $P_j^s \subseteq P_j$;

P_j^{ns}: sub-set of non-divisible care operations for the patient P_j, with $P_j^{ns} = P_j \setminus P_j^s$;

B: set of couples (m_1, m_2), where m_1 and m_2 are two members of the medical staff who have patients in common;

w_{kl}: number of patients in common to be treated by members of the medical staff m_k and m_l;

r_j : earliest starting date for treating patient P_j;

$d_{i,j,k}$: theoretical execution time for operation $O_{i,j}$ performed by m_k medical staff;

W_k : workload for m_k medical staff;

W: workload for all members of medical staff, $W = \sum_{k=1}^{M} W_k$;

$r_{i,j,k}$: earliest availability time for m_k medical staff for operation $O_{i,j}$;

$r_{i,j,k} \geq r_j$;

$C_{i,j,k}$: necessary competence of the medical staff m_k for performing the execution of operation $O_{i,j}$;

t_j^{ent} : P_j , patient's arrival date;

$t_{i,j}$: starting date for executing operation $O_{i,j}$;

$\gamma_{i,j}$: minimum theoretical duration for executing care operation $O_{i,j}$;

c_j: theoretical ending date for the treatment of the patient P_j;

d_j: real ending time for the treatment of the patient P_j.

5.2.1.2. Parameters associated with the care room

S: total of sites at CHRU of Lille (radiology, IMR, site for biological testing, etc.) and $n_S = Card(S)$ is the total number of sites in this set;

R: set of all the available care rooms;

n_p^R : number of available care rooms during the period p;

S_r^R : capacity of care room r, $r \in R$.

5.2.2. *Variables*

5.2.2.1. *Primary decision variables*

X_{jpr}: Boolean, equal to 1 if an operation or a part of this operation corresponding to the patient P_j is placed during the period p at care room r;

X_{jp}: Boolean, equal to 1 if an operation corresponding to the patient P_j is placed during the period p.

5.2.2.2. *Secondary decision variables*

A_{pr}: a whole representing the number of patients with divisible care operations during the period p at care room r;

C_{lk}^{xy}: Boolean, equal to 1 if members from the medical staff m_l and m_k, having patients in common, are placed at two consecutive periods x and y during the same day;

C_{lk}^{SP}: Boolean, equal to 1 if members from the medical staff m_l and m_k, having patients in common, are placed at two different sites S and P, in two periods, with a time gap;

U_{pr}: Boolean, equal to 1 if many patients are treated during the period p, at care room r. The number of patients must not exceed the capacity S_r^R;

U_{jpc}: Boolean, equal to 1 if one of many members from the medical staff are allocated during the period p to the patient P_j in corridor c, with a time gap;

C_{lk}^{T}: Boolean, equal to 1 if members from the medical staff m_l and m_k have patients in common and are placed at two periods with a time gap, and care rooms are placed at different sites. For this, a trip T is necessary.

5.2.3. *Institutional parameters*

w^T: penalty associated with the movement of one patient among the different sites of the health establishment;

w_c: penalty for using corridor c;

BC_p^r: penalty associated with exceeding the capacity of room r during the period p;

MS_p^k: penalty associated with exceeding the workload of the medical staff member m_k during the period p;

G^{PS}: penalty associated with the gap during the treatment period between sites S and P.

5.2.4. *The objective function*

Minimize:

$$C\left(w^T\right) + C\left(w_c\right) + C(BC_p^r) + C(MS_p^k) + C\left(G^{PS}\right) \qquad [5.1]$$

with:

$C\left(w^T\right)$: incurred cost due to transporting the patient along different sites at the healthcare center;

$C\left(w_c\right)$: incurred cost due to corridor use;

$C\left(BC_p^r\right)$: incurred cost for exceeding the capacity of room r during the period p;

$C\left(MS_p^k\right)$: incurred cost for exceeding the workload of the medical staff member m_k during the period p;

$C\left(G^{PS}\right)$: incurred cost due to gap in the treatment period.

The objective function is a sum of penalties, each of which features a specific subtle constraint.

5.2.5. *The constraints*

5.2.5.1. *Hard constraints*

The following hard constraints condition the feasibility of the solution:

SSO_r^p: The sum of divisible operations (or operation parts) allocated to the care room r during the period p must not exceed the room's capacity:

$$\forall r \in R, \forall p \in \mathbb{R}, A_{\mathrm{pr}} \leq S_r^R$$

5.2.5.2. *Subtle constraints*

The quality of the solution is determined by the following subtle constraints:

CPP_{lk}^{SP}: Each time two members of medical staff m_k and m_l with patients in common are placed during two consecutive periods at two different sites S and P, a patient movement penalty is applied:

$$C^{SP} = w^T \sum_{m_l, m_k \in MS} C_{lk}^{SP}$$

MPC_p: Each time a medical staff member is allocated to treat one or more patients in the corridor during the period p, a corridor penalty is applied:

$$U^c = W^c \sum_{P_j \in NP, p \in \mathbb{R}} U_{jpc}$$

Cap_p^r: Each time at least two patients are treated during the same period p at the same room r, a penalty for exceeding room capacity is applied:

$$BC^r = BC_p^r \sum_{P_j \in NP, r \in R, p \in \mathbb{R}} U_{jpc}$$

5.2.6. *Criteria*

5.2.6.1. *Reduction of the total workload of medical staff members: Cr₁*

Total workload equals the sum of the duration of all care operations according to a certain allocation of medical staff. However, for each $O_{i,j}$ operation, the execution duration is greater than minimum length $\gamma_{i,j}$, where

$$\gamma_{i,j} = \min_k \left(d_{i,j,k} \right).$$

Then, $Cr_1 \geq \sum_j \sum_i \gamma_{i,j}$

5.2.6.2. *Reduction of the global patient waiting time: Cr₂*

This corresponds to minimizing Cr₂, such that $Cr_2 \geq \sum_{j=1}^{N} \max(0, c_j - d_j)$.

5.2.6.3. *Reduction of reply response time for care tasks: Cr₃*

This corresponds to minimizing Cr_3, with: $Cr_3 \geq \max_j \left(r_j + \sum_i \gamma_{i,j} \right)$.

5.2.6.4. *Balancing medical staff workload: Cr₄*

Given the S allocation of a care operation to an MS member, and given $\overline{W}(S)$ as the average workload for a member of the MS following S allocation.

For the same S allocation, the workload of the medical staff (MS) member who has the highest workload is greater than the average workload, hence $\forall S$, and we have:

$$Cr_4(S) = \max_k \left(W_k(S) \right) \geq \overline{W}(S)$$

$\overline{W}(S) = \dfrac{Cr_1(S)}{M}$, with the aid of the lower bound of the criterion Cr_1, we

have: $\forall S, \ Cr_4(S) \geq \dfrac{\sum_j \sum_i \gamma_{i,j}}{Card(MS)}$

5.3. Multiple competence task

The problem to be solved is ensuring a quality service of patients coming to PES, while respecting the degree of the emergency. The purpose is to better manage patient flow, prioritizing the most serious cases. The management system of PES must mobilize members of medical staff according to their availabilities and competencies. The problem comes down to organizing the execution of care tasks while considering the competencies. Each treatment of the patient P_j corresponds to the execution of an n_j number of care operations. Each $O_{i,j}$ operation can be performed by a medical staff member with a percentage of the necessary competence for the task execution. Each patient P_j needs a variety of competencies for his treatment, simultaneously and/or in sequences.

Each medical staff member m_k has a competence level defined by an $\theta_{i,j,k}$ percentage, necessary for the execution of operation $O_{i,j}$. If $\theta_{i,j,k} = 100\%$, then $C_{i,j,k}$ is his main competence. Otherwise, $0 < \theta_{i,j,k} < 100\%$, $C_{i,j,k}$ is his secondary competence. In this case, m_k cannot take part in the execution of the care operation unless he is considered sufficiently qualified by the head of the medical team to which he belongs. If medical staff m_k does not have the minimum of required competence for performing care task $O_{i,j}$, he must be accompanied by the medical staff mastering this task for supervision. Finally, if $\theta_{i,j,k} = 0$ then the medical staff member m_k is incapable of performing this task.

Allocating a care operation $O_{i,j}$ to a member of medical staff m_k means that he will not be available during $d_{i,j,k}$ $(d_{i,j,k} \in N^*)$. During this time period, this staff member is not available for executing other tasks.

Competence examples:

Competence Id	Description
C_1	Injection
C_2	Intubation
C_3	Surgical implementation

Table 5.1. *Example of medical competencies*

If we identify C_k, $k \in \{1, 2, 3\}$, each mastered competence by each m_k actor of the PES medical team, the consideration of the polyvalence of these actors can be manifested, for each of them, by the existence of a knowledge degree belonging to [0,1] for each m_k person and for each required competence.

A degree equal to 1 identifies the principal or main competence of an actor (specialization). If the knowledge degree is non-null and different from 1, the actor in consideration can take part as an assistant or replace more qualified actors. The example below (Tables 5.1 and 5.2) reflects the fact that medical actor m_1 masters competencies C_1, C_2 and C_3 with respective degrees equal to 1; 0.9 and 0.7. Therefore, C_1 is his main competence or, in other terms, his specialization. C_2 and C_3 are equally acquired competencies for m_1.

Competencies	Staff			
	m_1	m_2	m_3	m_4
C_1	1	0.4	0.4	0
C_2	0.9	1	1	0.6
C_3	0.7	1	1	1

Table 5.2. *Example of staff experience level table*

The latest P_j patient can be treated on the deadline date. The order of care operations for the patient's treatment time is fixed from the start thanks to medical protocols. The example of identification of care operations provided in Table 5.3 expresses that operation 1 for patient 1 ($O_{1,1}$) with a deadline set on 12, requires 2 time units for his treatment with competencies C_1 and C_2. In this case, the medical team is formed by two persons mastering these 2 competencies, for example, a doctor and a nurse.

Competencies	Operations (Limit intervention deadline)						
	$O_{1,1}(12)$	$O_{2,1}(14)$	$O_{1,2}(5)$	$O_{2,2}(7)$	$O_{3,2}(9)$	$O_{1,3}(10)$	$O_{2,3}(11)$
C_1	2	0	2	2	0	2	0
C_2	2	2	2	2	1	2	2
C_3	0	2	0	2	1	0	2

Table 5.3. *Example of identification table of care operations requiring multiple competencies*

The mathematical formulations introduced here play a theoretical base role for scheduling care tasks in collaborative Workflow in an optimal way. Taking into account the distributive nature of collaborative Workflow as suggested in Chapter 4, unforeseeable events may appear at any moment (emergencies, epidemics, etc.). Thus, the coordination of task execution by agents must be done in a dynamic way in order to neutralize or reduce the impact of disturbances. An architecture modeling the resolution of patient care at PES must integrate coordination, collaboration, scheduling and orchestration approaches. For this reason, in the following section, we propose a multi-agent architecture capable of taking into account all these useful aspects at PES.

5.4. Agent-based modeling

Thanks to a top-down approach, we have conceived a model composed of various types of agents: Reception Agent (ReA), Problem Identification Agent (PIA), SA, Tracing Agent (TA), Resource Agent (RA), Medical Staff Agent (MSA), and Integration and Evaluation Agent (IEA) [BEN 14b].

On arrival at Emergency Services, the first contact established is with the Reception and Orientation Nurse (RON). The role of the nurse is to welcome the patient and to obtain the necessary information about his health condition, medical history, etc. In general, an administrative reception takes place before being received by the RON, whose role is to register the patient.

The second role of the RON is to place the patient, depending on resource availability, directly in a box or a waiting room until treatment by a doctor. Other complementary medical examinations may be necessary throughout the treatment process. The actors of PES are modeled by agents who are autonomous, intelligent, active, dynamic and cognitive. They can have different roles to ensure that all care tasks are performed by the medical staff members (Figure 5.1).

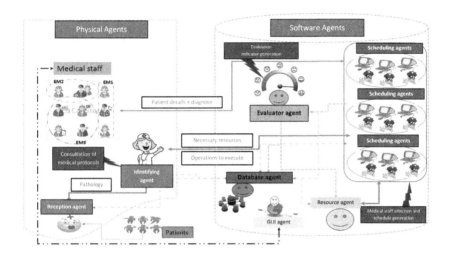

Figure 5.1. *MA proposed architecture*

Thanks to the information obtained during the interviews with the medical staff at the PES from the CHRU in Lille, two types of agents were identified: software agents and physical agents. Physical agents represent human actors, active on the ground (PES): patients, care staff (nurses and doctors), technicians, etc. On the other hand, software agents add functionalities to the suggested system, such as a patient information system, resource management and the calculation of performance indicators. In the following section, we will detail the suggested multi-agent architecture. In this architecture, there are several types of functional agents: those for patient reception and orientation, as well as for identifying pathologies. After registration, he sends the patient's information to the TA in charge of monitoring the patient's condition and localization, as well as to the PIA, who defines the necessary resources for treating each patient as well as the required competencies for each care task. The PIA then notifies with RA and the SA the information concerning the human and material resources to be allocated.

During the process of human resource allocation, the SA communicates with the MSA in order to identify the medical staff member capable of providing care to a specific patient. This is associated with their competencies and availabilities. Each medical staff member is modeled by a MSA, which is a mobile agent. He can intelligently move from one medical team to another in order to treat patients. During the execution of planning, patients with a high degree of emergency have priority over others. Once planning has been executed, the IEA generates performance indicators to globally evaluate the generated schedule.

In the following section, we will detail the behavior of different agents involved in the proposed architecture.

5.4.1. *Reception Agent's behavior*

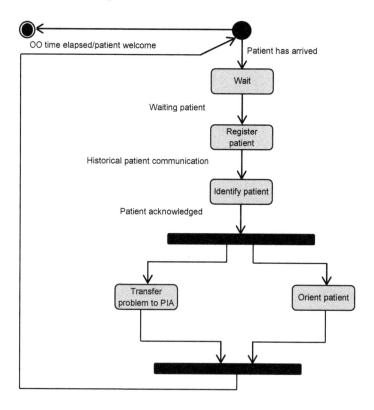

Figure 5.2. *RA's behavior*

A patient arriving at PES for a consultation must be added to the hospital's database by the medical staff. That is the RA's task. The person responsible for welcoming patients registers the patient's Social Security number in the interface prepared for this. The RA then consults the medical history of a patient in the patient database in order to examine the drugs the patient is taking and informs the PIA of his medical history.

5.4.2. PIA's behavior

The behavior of the PIA is shown in Figure 5.3.

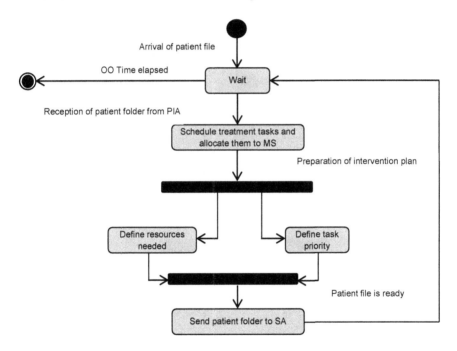

Figure 5.3. *PIA's behavior*

The PIA receives all the information related to the RA's patient. Then, he receives the patient's updated medical history, integrating the diagnosis made by the doctor who treated him. This agent acts as a "regulating doctor". He generates the treatment plan for the patient, creating a list of care tasks to be performed. He also determines the required medical staff for each care task. Finally, he registers all this information in the patient's medical history and transmits these data to the SA, who allocates care tasks to the different members of the medical staff, determining their execution starting dates.

5.4.3. *TA's behavior*

The behavior of the TA is represented in Figure 5.4. He mainly takes part when the execution plan is modified, be it because a new task has been added or because a planned task has been suppressed. The members of medical staff inform the TA about the modifications to be applied and the TA transfers this information to the PIA. He then interacts with software agents and informs physical agents about the actions and the health condition of patients.

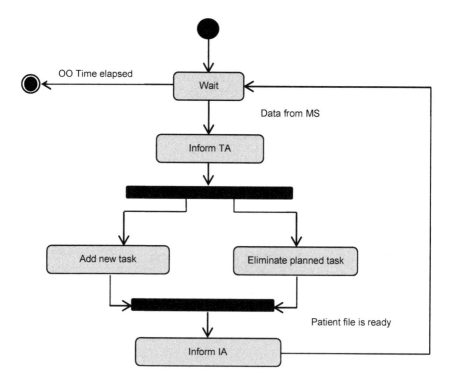

Figure 5.4. *TA's behavior*

5.4.4. *Resource Agent's behavior*

The behavior of the Resource Agent (RA) is shown in Figure 5.5. This agent is in charge of following up and managing different available human and material resources for care tasks. He updates the available date of medical staff members, monitors medicine stock and manages inventories.

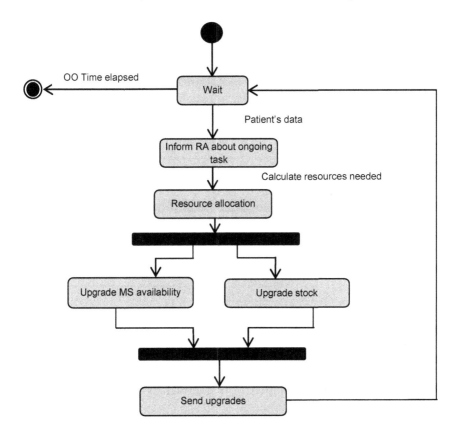

Figure 5.5. *ReA's behavior*

5.4.5. *IEA's behavior*

The behavior of the IEA is shown in Figure 5.6.

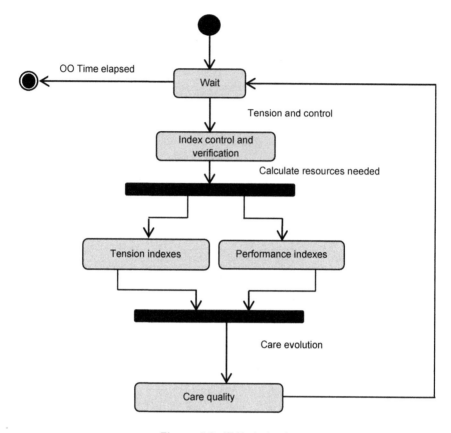

Figure 5.6. *IEA's behavior*

This agent is in charge of evaluating the system's performance. He verifies and controls the evolution of overcrowding indicators. He equally calculates the system's performance indicators such as patient's waiting time, in order to evaluate the quality of patient care at PES.

5.4.6. *MSA's behavior*

The behavior of the MSA is shown in Figure 5.7.

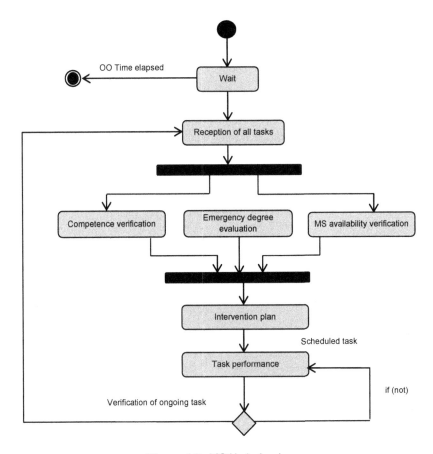

Figure 5.7. *MSA's behavior*

A MSA is a mobile agent, who physically represents a medical staff member and can move from one medical team to another at PES in order to treat patients. He is characterized by two variables (competencies and availability). One care task can belong to one or more patients. The MSA may simultaneously receive many treatment demands from different patients and it is according to their availability and emergency condition that the patients will follow the treatment order in different boxes. This particular type of agent is equipped with data, conditions, a code and an intelligent behavior. Once the MSA performs a care task on a patient, he may move on to another team to execute a new care task for another patient. As a consequence, the SA must take this mobility aspect into account when allocating human resources to tasks. Each care task can be performed by different possible MSAs, with different durations, which depend on the level of experience of each medical staff member.

5.4.7. *SA's behavior*

The behavior of the SA is shown in Figure 5.8. This agent has to optimize the scheduling of diverse care tasks for various patients, taking into account different system constraints. He has to allocate resources to care tasks minimizing patient's waiting time.

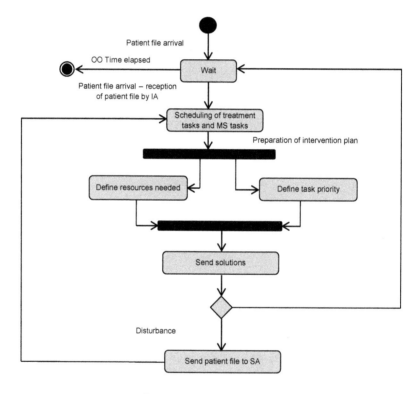

Figure 5.8. *SA's behavior*

Having introduced the behavior of each agent in our system, we shall focus on the integration of EA[1]-based Metaheuristics into the behavior of

---

1 NT: EA.

the SA, in order to optimize the scheduling of care tasks and smooth activity peaks during overcrowding periods.

5.5. Description of a SA's behavior

The SA uses different variables, constraints and criteria detailed in section 5.2 to provide an optimal and feasible care task schedule. The optimization method we have integrated into the behavior of the SA is based on Evolutionary Algorithms (EA). The SA is incapable of scheduling multiple competence care tasks on his own. Apart from the EA he is equipped with, he needs to interact with different agents from the proposed architecture so as to coordinate their activities and their behavior and accomplish a common goal: to provide an optimal schedule of care activities and to ensure quality care to patients [BEN 15a]. A schematic of this interaction is shown in Figure 5.9.

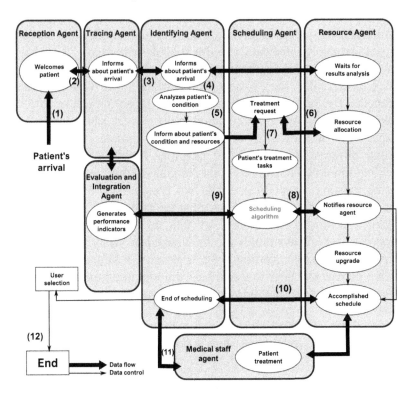

Figure 5.9. *SA's different interactions with other system agents*

5.5.1. *Different phases in the behavior of the SA*

Scheduling multiple competence care tasks is an NP-hard combinatory optimization problem. Therefore, it is more sensible to use Metaheuristics based on EA, which make it possible to give a set of good approach solutions (not necessarily optimal), at every moment [BEN 15d]. Potential solutions provided by EA imperatively need the coordination of different system agents for their exploitation.

The behavior of the SA is based on two phases, depending on the size of the problem at hand: phase 1 for small- and medium-sized problems (example: $1 \leq N \leq 20$ and $\forall j, 1 \leq n_j \leq 5$) and phase 2 for large-sized problems (example: $N \geq 20$ and $\forall j, n_j \geq 5$).

5.5.1.1. *SA: Phase 1*

During the first phase, a task must be inserted at a reception interval in the schedule. It is a time interval available to receive the totality of the task without interfering with neighboring tasks. For this, the SA first searches in the list of available medical staff for people who have the necessary competence to perform the operation that we want to insert in this reception interval. Consequently, the SA sets up a medical team with the necessary competencies for performing this operation. If no reception interval is available for receiving the care operation in the current schedule, the SA can make space for an extension of the most appropriate reception interval (respecting different constraints and priorities introduced in section 5.2), rescheduling top-down operations corresponding to this reception interval. If there is no existing reception interval, two possibilities may appear: if the operation is not urgent, it will be placed at the end of the sequence. If it is, non-priority operations will be left on hold, to make way for this urgent operation. The SA always builds the final schedule verifying its coherence. However, if the chosen starting time is around 3 hours behind the last patient, the patient will have to wait. Such waiting time is necessary because, for example, the study of biological samples often takes longer than 2 hours.

After having allocated a task to each medical staff member, for adaptation questions concerning their working rhythm, the SA imposes a 20-minute rest time if, at the end of the task, the last pause goes back to 2 hours earlier.

5.5.1.2. *SA: Phase 2*

Phase 2 of the SA's behavior is dedicated to bigger problems. In this context, we have chosen metaheuristics based on EA for finding a set of acceptable solutions, knowing that a solution is modeled by a chromosome, for multiple competence scheduling task problems.

In this phase, allocation and scheduling of care operations are made jointly.

5.5.2. *Allocation algorithm*

This enables allocating each care operation to the appropriate medical staff member, taking into account their availability and workload. A simple application of flow logic is proposed in order to calculate the available time of each medical staff member m_k. This calculation is based on the analysis of competencies, the evolution of the ongoing care task, the evolution of the patient's condition and the seriousness of the pathology. These are the flow entries. Then, we define three sub-flows for each entry {"weak", "average" and "High"}. Each sub-flow is characterized by its trapezoidal Belonging Functions (BF), which depend on the variability of the entries. The definition of the variables' BF or inference is based on decision rules as a function of expert's opinions and historical data [BEN15c].

Examples of rules:

If (the medical staff member is "highly qualified") && (the evolution of the ongoing care deed is "high") && (the pathology is "serious") && (the patient's condition "improves"), then (the medical staff member is "highly available").

The result is that a fuzzy value is de-fuzzed in order to obtain an exact number (as final exit), using the gravity center method. For allocation, we choose to allocate a care operation to the medical staff member which corresponds to the fuzziest value, which reflects his availability rate. If two medical staff members have the same fuzzy value, we choose to balance the workload among all the medical staff members. This allocation procedure allows us to build an E allocation set $(E = \{S^c \,/\, 1 \leq z \leq cardinal(E)\})$ and to balance the workload of medical staff. Each allocation is represented in S^c, $S^c = \{S^c_{i,j,k} \,/\, 1 \leq j \leq N; \, 1 \leq i \leq n_j; \, 1 \leq k \leq M\}$ table. For each i, j and k, the value of $S^{c}_{i,j,k}$ can be between 0 and 1. The value "$S^{c}_{i,j,k}=1$" means that $O_{i,j}$ is allocated to m_{k}. The value "$S^{c}_{i,j,k}= 0$" means that $O_{i,j}$ cannot be allocated to m_{k}.

id	designation	glasscow	hydration	temperature	heart rate	bloodpressure	respiratoryfrequency	cerebralscanner	bacteriologicaltest	radiothorax	neurologicaltest	fracturescore	bloodtest	skeletonradio
1	concussion	1	0	1	1	1	1	3	0	3	2	0	2	0
2	gastroenteritis	0	1	1	1	1	1	0	2	3	0	0	1	0
3	compound fracture	0	0	1	1	1	1	0	0	0	0	1	2	1

Table 5.4. *Medical staff competencies*

A possible scheduling plan as a function of medical staff competencies (Table 5.4) is given in Table 5.5. We consider that the allocation of a $O_{i,j}$ care operation to a medical staff member is possible when the competence $C_{i,j,k} \geq 0.5$.

$S^{c} = \{ S^{c}_{i,j,k} / 1 \leq j \leq N ; 1 \leq i \leq n_j ; 1 \leq k \leq M\}$		m_1	m_2	m_3	m_4
Chapitre I	$O_{1,1}$	0	*	*	0
P_1	$O_{2,1}$	0	1	0	0
	$O_{3,1}$	0	*	*	*
Chapitre II	$O_{1,2}$	*	0	0	*
P_2	$O_{2,2}$	0	0	1	0
	$O_{3,2}$	1	0	0	0
Chapitre III	$O_{1,3}$	*	*	0	*
P_3	$O_{2,3}$	*	*	*	0

Table 5.5. *Generation of an S^z allocation diagram*

The value "$S^c_{i,j,k} = 0$" shows that the medical member staff m_k is not qualified for this operation. The value "$S^c_{i,j,k} = 1$" shows that the allocation of $O_{i,j}$ operation to the medical staff member m_k is obligatory because he is the only one for whom $C_{i,j,k} \geq 0.5$. In this case, we complete the line with "0". Symbols "*" show that allocation is possible $(C_{i,j,k} \geq 0.5)$.

5.5.3. Evolutionary algorithm

The allocation diagram introduced in the previous section becomes the chromosome model for our EA. This chromosome adapts easily to the properties and constraints defined in section 5.2 and is useful for the construction of individuals in order to integrate the right properties and the good execution of medical protocols. With a valid chromosome diagram, EAs are faster and more efficient, building solutions which favor the reproduction of individuals that respect the right diagrams. In the case of the scheduling problem we are analyzing, the difficulty in implementing this technique is important because it needs to elaborate a very particular code, which can simultaneously describe the problem's data and exploit the diagram's theory.

Table 5.6 presents an example of a chromosome diagram featuring $\{0, 1, *\}$.

III.1		m_1	m_2	m_3	m_4
Chapitre IV	$O_{1,1}$	0	*	*	0
P_1	$O_{2,1}$	0	1	0	0
	$O_{3,1}$	0	*	*	*
Chapitre V	$O_{1,2}$	*	0	0	*
P_2	$O_{2,2}$	0	0	1	0
	$O_{3,2}$	1	0	0	0
Chapitre VI	$O_{1,3}$	*	*	0	*
P_3	$O_{2,3}$	*	*	*	0

Table 5.6. *Chromosome example*

This chromosome diagram covers all the interesting possibilities for allocating care operations. Nevertheless, it presents bans which can be costly in terms of medical staff workload. However, allocations belonging to this model do not guarantee an optimal solution in terms of workload distribution. Certain allocations can be left aside by the SA when they do not provide satisfying solutions. Indeed, during the reproduction phase, if the individual does not respect the chromosome's diagram, it is instantly rejected. Such an approach makes it possible to restrict searching space, which will accelerate the algorithm's convergence and ensure good quality solutions.

5.5.4. Crossover operations

These operations are useful for exploring research space and for diversifying population, thanks to the exchange of information between two individuals, through the use of a mask. The crossover algorithm is as follows:

Crossover Algorithm
– Randomly choose 2 parents S^1 and S^2;
– Randomly choose 2 whole numbers j and j' such that $j \leq j' \leq N$;
– Randomly choose 2 whole numbers i and i' such that $i \leq n_j$ and $i' \leq n_{j'}$ (in the case where $j=j'$, $i \leq i'$);
– The allocation in f^1 must correspond to the same allocation in S^1 for all the operations in lines (i,j) and (i',j');
– The rest of the allocation in f^1 must correspond to the same allocation in S^2;
– The allocation in f^2 must correspond to the same allocation in S^2 for all the operations on lines (i,j) and (i',j');
– The rest of the allocation in f^2 must correspond to the allocation in S^1;
– Update of execution starting dates;
VI.1.1

EXAMPLE (Figure 5.10).–

Let us suppose that S^1 and S^2 have been chosen and that $j=1$, $j'=2$, $i=2$, $i'=2$. f^1 and f^2 have been generated applying the crossover algorithm.

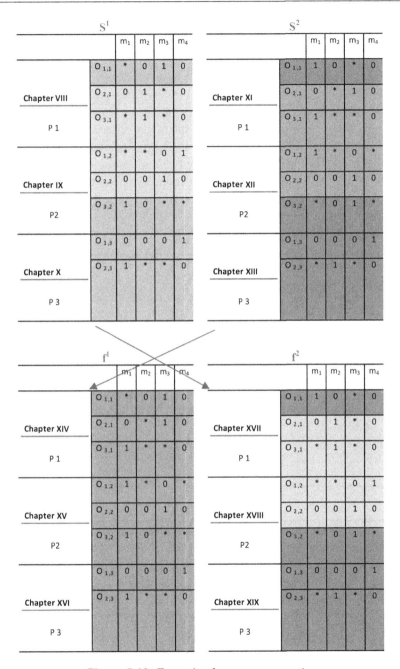

Figure 5.10. *Example of crossover operator*

5.5.5. *Mutation operators*

Mutation is an asexual operator, which needs only one chromosome in order to generate a child chromosome. These operators make it possible to maintain the random aspect in the evolution of the population in order to avoid premature convergence. In our case, mutation operators are controlled in order to favor the improvement of considered criteria. For this, we have implemented many mutation operators: controlled scheduling mutation, controlled allocation mutation, controlled mutation balancing the workload of medical staff and controlled mutation reducing the entry time of a patient to PES.

Below, 2 mutation operator algorithms:

Controlled mutation reducing patient treatment duration at PES

– *Randomly choose an S individual;*

– *Choose a care task corresponding to a j patient with the broadest Cmax: (Max $_j$ {C_{maxj} with $C_{maxj} = \sum_i \sum_k S_{i,j,k} d_{i,j,k}$});*

– *i=1; r = 0;*

– **As long as** *(i≤n_j et r = 0)*

   - find k_0 such that $S_{i,j,K0} = 1$;

   *- **For** (k=1, k≤M)*

        *If (d $_{i,j,k}$ < d $_{i,j,k0}$) **then** {$S_{i,j,K0}$ =0; $S_{i,j,K}$ =1; r=1 ;}*

        End if

      End for

   - i=i+1;

End

– *Update execution starting dates*

Controlled mutation balancing medical staff workload

– *Randomly choose an S individual;*

– *Find medical staff member with the highest M_{k1} workload (Max$_k$ {W_k /$W_k = \sum_j \sum_i S_{i,j,k} d_{i,j,k}$});*

– *Find medical staff member with the lowest M_{k2} workload (Min $_k$ {W_k});*

– *Randomly choose an operation $O_{i,j}$ such that S $_{i,j,k1}$ =1;*

– *Allocate this operation to medical staff member with the lowest workload:*

   $S_{i,j,k1}$=0;

   S $_{i,j,k2}$=1;

– *Update execution starting dates*

5.5.6. *Selection operators*

After crossover, the population size increases. In fact, we add children algorithms to their parent algorithms. It is then necessary to choose the chromosomes that will become a part of the new population.

For this, all the chromosomes must be evaluated. For each solution, we have to calculate its fitness and normalize it (put it in a percentage form of the total force). Choosing the strongest solutions does not necessarily guarantee enough diversity of solutions in the population, and choosing randomly could possibly deprive us from good solutions. We have therefore chosen to keep only a percentage of the best solutions, then to choose those that remain at the wheel (their selection probability corresponds to their normalized force). Thus, we ensure a strong and varied selection.

In the following section, we introduce the chosen method for evaluating fitness function.

5.6. Dynamic aggregative approach for evaluating fitness function

We use an aggregative approach with dynamic weight in order to evaluate the quality of solutions:

$$F_{fitness}(x) = \sum_{q=1}^{q=L} w_q \cdot f_q(x)$$, where w_q is the weight of the qth objective function, L is the total number of criteria and $w_q \in [0,1] \ \forall \ 1 \leq q \leq L$. The weight sum equals 1.

The calculation of the global fitness function is based on the dynamic generation of weight $w_q (1 \leq q \leq L)$ by flow logic

$$f_g(x) = \sum_{q=1}^{q=L} w_q \cdot \mu_q^G(f_q(x))$$, where $\mu_q^G(\ )$ is the membership function of a criterion that has a fuzzy sub-set.

The purpose is to measure the average quality of solutions at each iteration k, according to each criterion and its lower bound.

Given $\bar{f}{}^{k}_{q}$ the average value of all criteria for one Pop_k population,

solutions are: $\bar{f}{}^{k}_{q} = \dfrac{\sum\limits_{x \in P_k} f_q^k(x)}{cardinal\,(Pop_k)}$

Criteria evaluation is based on membership functions (Figure 5.11). The two fuzzy sub-sets are thus defined:

– $Near_q$: the sub-set of solutions near lower bound f_q^* according to the qth objective function;

– Far_q : the sub-set of solutions far from lower bound f_q^* according to the qth objective function.

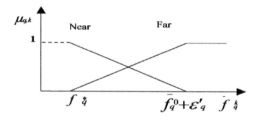

Figure 5.11. *Membership functions according to objective functions*

At each k iteration, membership functions are formulated as follows:

$$\mu^{Far}_{q,k}\,(\bar{f}{}^{k}_{q}) = \frac{\bar{f}{}^{k}_{q} - f_q^*}{\bar{f}{}^{0}_{q} - f_q^* + \varepsilon'_q} \quad \text{if } \bar{f}{}^{k}_{q} \in \left[f_q^*,\, \bar{f}{}^{0}_{q} + \varepsilon'_q \right];\ \text{if not}$$

$$\mu^{Far}_{q,k}\,(\bar{f}{}^{k}_{q}) = 0$$

or ε'_q a small positive value to solve the zero division problem (when $\bar{f}{}^{0}_{q} = f_q^*$). $\varepsilon'_q = 0.01 \cdot f_q^*$ if $\bar{f}{}^{0}_{q} = f_q^*$; if not $\varepsilon'_q = 0$

Weight calculation w_q^{k+1} is done applying fuzzy rules:

If ($\bar{f}_q^{\,k}$ is near f_q^*) then (w_q^{k+1} ↓),

If ($\bar{f}_q^{\,k}$ is far from f_q^*) then (w_q^{k+1} ↑).

Thus, $w_q^1 = \dfrac{1}{L}$ $\forall\, 1 \leq q \leq L$; $w_q^{k+1} = \dfrac{\mu_{q,k}^{Far}(\bar{f}_q^{\,k})}{\displaystyle\sum_{q=1}^{q=L}\mu_{q,k}^{Far}(\bar{f}_q^{\,k})}$ $\forall\, 1 \leq q \leq L$

and $\forall\, 1 \leq k \leq Q\text{-}1$ where Q is the number of iterations.

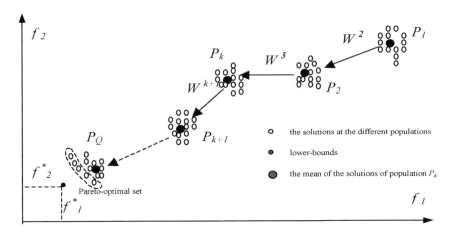

Figure 5.12. *Fuzzy dynamic control of research directions*

The vector of different weights (W^1, W^2,.., W^k,.., W^Q) is calculated in a dynamic way, evolving from the kth P_k generation to P_{k+1} generation, as a function of the distance between lower bounds and the average of individuals of P_k generation (shown with a black circle in Figure 5.12). The objective is to explore possible improvements to solutions. Priority is given to the optimization of objective functions whose average values are far from optimal (or the value at the lower bound). Consequently, while using this fuzz technique, it would be possible to control the research direction in order to build a final set of solutions near the optimal solution. This final set is used for choosing Pareto-optimal solutions and rejecting others.

This aggregative approach makes it possible to provide a scheduling solution that represents a good compromise between different criteria. The established schedule proposes a GANTT diagram to medical staff and a priority order between optimized operations. Despite the fact that this schedule has been optimized, it does not allow us to take into account the evolution of a patient's condition, particularly after a biological test, RMI, X-ray, etc. Consequently, for this schedule to be efficient, it must be coupled with a patient circuit dynamic orchestration approach in order to improve the ongoing criteria. We then need a global architecture for workflow dynamic orchestration, modeling the patient circuit based on the agents we will introduce in the following section, where we shall focus on the alliance between a multi-agent approach and Workflow, in order to achieve a dynamic orchestration of successful quality care at PES.

5.7. Workflow orchestration

5.7.1. *Definitions and concepts*

Computerized orchestration is based on an inherent intelligence with an implicitly autonomous control. It is used for generating coordination processes with information exchanges through the interaction of services, and corresponds to the automatization of systems deploying regulation elements. Orchestration thus makes it possible to organize a set of tasks, which require human and material resources, in a dynamic, collaborative and distributed manner. It provides the correct information at the right time, to the right resource, so that all tasks are performed efficiently, following a Workflow. Orchestration then describes the automatic process of organizing, coordinating and managing complex information systems.

5.7.2. *Health orchestration*

Health orchestration is an innovative way of optimizing patients' appointments in real time. Indeed, orchestration depends on the result of each task during the care process (example: X-ray, MRI, consultations, etc.).

This orchestration makes it possible to coordinate and manage processes and services modeled by Workflow. It is particularly important at ES because it enables a controlled execution of Workflow, modeling patient circuits at these services.

In the specialized literature, the works of Meli *et al.* [MEL 14] highlight the interest of care flow dynamic orchestration for reducing ongoing criteria Cr_1, Cr_2, Cr_3 and Cr_4, which comes down to reducing patient waiting time, regulating their rate of flow and smoothing the workload of medical staff. Besides, these analyses enable logisticians at healthcare establishments to reduce tension-associated costs and to have a better vision of available resources at PES. We can summarize the advantages of health orchestration as follows:

– coordination of different patient flows for significantly reducing care length and waiting time;

– smoothing up all care processes and frequent tasks involving multiple competencies: finding the right staff with the right competence at the right moment for the right patient;

– supervising human and material resources allocated to the patient, and informing the latter about the availability rate of these resources, as well as planned care protocols;

– allocating a care service to the right patient, being aware of resource availability;

– controlling and following up patient circuit, modifying their orientation in real time, depending on their health condition and care results.

5.7.3. *Dynamic orchestration architecture*

In order to contribute to the evolution of the previously introduced PES management system (formulation, workflow, agents, etc.), we suggest a three-layer architecture (Figure 5.13), which accepts the addition of new constraints, new criteria and decision variables. This architecture also accepts the addition of a software agent or a physical agent (e.g. the recruitment of a new nurse).

The purpose of this architecture is to identify a reference logistic situation, through mathematical and algorithmic models. In the case of disturbance (overcrowding) in the environment associated with the real PES logistic situation, the three-layer architecture enables us to quickly achieve the normal functioning of the service, thanks to the *collaborative optimization* of second-layer agents and workflow dynamic orchestration.

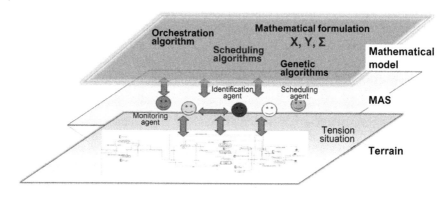

Figure 5.13. *Dynamic orchestration architecture*

The suggested dynamic orchestration architecture (Figure 5.13) consists of:

– *One first layer*, consisting of mathematical and algorithmic models (section 5.2) which make it possible to characterize a normal situation at PES.

– *One second layer* based on communicating agents (section 5.4), making it possible to identify a real ground situation. In the case of overcrowding, the normal situation at PES should be reached thanks to a collaborative optimization process among agents.

– *One third layer,* "ground", representing the situation of the real functioning of PES, modeled by workflow (Chapter 4).

This architecture, which is innovative and generic, is based on communicating agents representing the different actors in the supply chain, who are in direct relation with what is happening on the ground. These agents continuously scrutinize the information on the ground layer, comparing the real situation with the reference logistic situation. According to this information, as well as different available mathematical models (first layer), these agents will have to adapt their roles and their behavior in order to better react to the different ground disturbances, with the purpose of quickly achieving a reference logistic situation.

In this context, we choose the Workflow representing patient circuit at PES as our ground layer (Chapter 4). The "agent" layer of this architecture plays the key role of orchestra director, thanks to the interaction of its agents, in order to efficiently accompany the care of patients at PES.

5.7.4. Performance evaluation of dynamic orchestration

The evaluation of orchestration performances is measured by the satisfaction of criteria Cr_1, Cr_2, Cr_3 and Cr_4 during Workflow execution. In fact, each criterion is evaluated orchestrating workflow in a static mode (the case in which the order of tasks remains unchanged all throughout execution) and in a dynamic mode (in the opposite case). For each type of orchestration, we calculate the average value of each criterion for a certain number of patients during the simulation period.

Given the gain GCr_i corresponding to the criterion Cr_i for the dynamic orchestration of workflow WD (Cr_i) in comparison with a static orchestration WS (Cr_i), we obtain:

$$GCr_i = \frac{WD(Cr_i) - WS(Cr_i)}{|WS(Cr_i)|} \times 100$$

If $GCr_i < 0$, then Cr_i has been improved for a dynamic orchestration.

If $GCr_i > 0$, then Cr_i has deteriorated for a dynamic orchestration, in comparison with a static orchestration.

If $GCr_i = 0$, there is no modification in the values of considered criteria.

5.8. Simulation and results

5.8.1. *Choice of the multi-agent platform*

The need for implementing systems with many autonomous components demands a software infrastructure used as an environment for deploying and executing a set of agents. This infrastructure is called MAS[2] development platform.

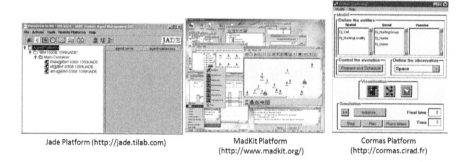

Jade Platform (http://jade.tilab.com)   MadKit Platform   Cormas Platform
                                             (http://www.madkit.org/)   (http://cormas.cirad.fr)

Figure 5.14. *Multi-agent platforms*

Nevertheless, the implementation of such systems is often difficult for manipulating complex data structures, distribution, communication and imposed material constraints. Besides, Artificial Intelligence (AI) is an extremely rich field of research and this richness implies a great complexity and multiplicity of approaches, which very often leads to numerous models of agents, environments, interactions and organizations. These models are often combined at the core of one MAS. Thus, the best option is to choose a multi-agent platform adapted to the constraints of the system to be implemented. Many multi-agent platforms exist, such as MadKit, JADE, ZEUS, AgentBuilder, Jack, etc. (Figure 5.14).

For platform selection, insignificant criteria such as learning difficulty or lack of resources were not considered. However, certain important criteria are to be highlighted[3]:

– the possibility of implementing relatively complex systems;

---

2 NT: (MAS) Multi-agent System.
3 http://jade.tilab.com.

– flexibility: avoiding platforms which support a particular methodology;

– development acceleration, thanks to the sufficiently important presence of software bricks in order to produce an accomplished application;

– distributed treatment, particularly the presence of AM paradigm support;

– support for mobile devices and web applications;

– possibility of integrating web services;

– graphic execution environment;

– application in compliance with FIPA regulations;

– open source;

– possible development of new interaction protocols.

The two platforms which do not specify any methodology and can be considered as "Framework[4]" are JADE and Jack, but JADE is superior because it has many interesting features such as the possibility of integrating web services and a good support for content languages and ontologies. Due to this, we have chosen the JADE (Java Agent Development Framework) platform for developing our proposed system and the results of our optimization approaches.

JADE is a "middleware[5] "software that enables a flexible implementation of communicating MAS thanks to the efficient transfer of ACL (Agent Communication Language) messages, in compliance with FIPA specifications. JADE is written in Java, supports mobility, is in rapid evolution and is today one of the few multi-agent platforms that offer the possibility of integrating web services [GRE 05]. On the other hand, JADE intends to facilitate the development of agent applications by optimizing the performances of a distributed agent system.

---

4 A set of libraries enabling the rapid development of applications. It provides enough software bricks in order to produce an application.

5 Enables communication between *customers* and *servers* with different structures and implementation procedures.

Despite the fact that the JADE project was conceived more for academic rather than industrial reasons, and even if the platform has been available on Open Source since May, 2003 (under Lesser General Public Licence Version 2), JADE has been used in numerous international projects[6]:

– Project: Multi-agent Scheduler, distributed by *ERXA* (*Engineering, Robotics & Control System Applications*), consists of the creation of an industrial planner for manufacturing environments. *ERXA* uses JADE for developing this project, which is called MASP – *Multi-Scheduler Project Agent*, and aims to create a generic scheduler based on the multi-agent technique.

– *Fraunhofer IITB (Fraunhofer Institut für Informations- und Datenverarbeitung)* uses JADE for the development of a great variety of agent-based software systems. In the field of manufacturer control, Fraunhofer has implemented *ProVis.Agent* – a system using JADE for real-time automobile industry production control.

5.8.2. *Functionalities and configuration*

In order to catch a glimpse of agent communication and behavior, JADE offers graphic tools – which are agents themselves – such as the "RMA" (Remote Management Agent), which represents the principal management interface (Figure 5.15), the "introspector", which can control the lifecycle of an agent, his exchanged ACL messages and behavior, and who can trace down the progress of communication among different system agents (Figure 5.16).

There is a final important tool on the JADE platform, which corresponds to the *"Directory Facilitator Agent"*, or *"DF"*. It is a yellow page service for indirect communication managed by the DF. In this way, every agent providing a service can announce himself before the DF, and every agent looking for a service can consult the directory.

---

6 http://en.wikipedia.org/wiki/Comparison_of_agent-based_modeling_software.

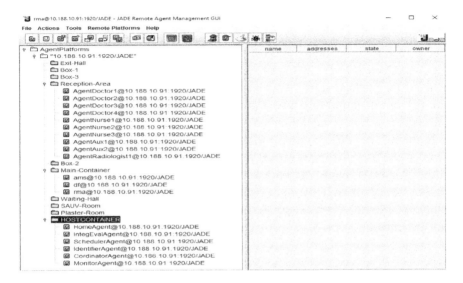

Figure 5.15. *"RMA"*

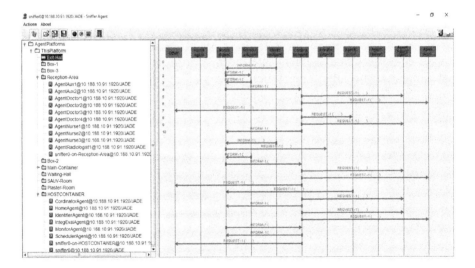

Figure 5.16. *Agent communication within the system*

5.8.3. *Developed dynamic orchestration system*

Our system, called SysCAOO (*Système Collaboratif d'Aide à l'Ordonnancement et l'Orchestration*[7]), is a decision-support system intended for medical staff members for managing and following up patients, with the aim of improving care quality. It helps PES staff to make decisions concerning patient orientation and treatment as well as to evaluate the performance of PES. Indeed, thanks to SysCAOO, each medical staff member can access the necessary information about the number of patients present at PES, under treatment or awaiting treatment. They can also monitor the evolution of care. SysCAOO offers better coordination and communication at the heart of medical teams and between these teams and their patients.

SysCAOO is a flexible, evolving and interactive tool. It has many ergonomic interfaces, which enable PES administrators to take part and act through the system's functionalities.

5.8.4. *Interface implementation of the developed system*

At the start of simulation, SysCAOO loads real-life PES databases. Our system comprises management interfaces such as medical staff registration and authentication interfaces. When we launch the system, user registration and/or authentication interfaces appear. This step is important for securing patient data (Figure 5.17).

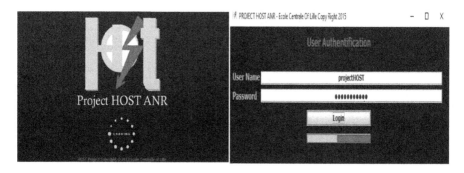

Figure 5.17. *Authentication interface*

---

7 NT: Collaborative System for Scheduling and Orchestration Support.

5.8.4.1. *Main interface*

The main interface, shown in Figure 5.18, allows us to visualize different transactions and interactions between the different agents in our system.

Figure 5.18. *Main interface*

5.8.4.2. *Theater interface (PES)*

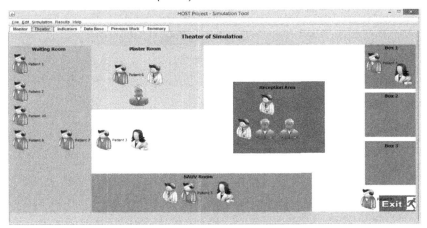

Figure 5.19. *Theater interface (PES)*

The theater interface, shown in Figure 5.19, emulates the architecture of PES, including boxes, waiting rooms, etc. During task execution, it helps us follow patient circuits as well as medical staff movements in the service (all these are modeled by mobile agents in our multi-agent proposed architecture).

5.8.4.3. *Criteria evolution interface*

The criteria evolution interface, shown in Figure 5.20, is used for following the changes in medical staff availability and patient circuit. The board on the left refers to the follow-up of patient care. For example, the moment a patient arrives, the waiting time for the first service is shown by a red indicator, and the corresponding progress bar is empty (0%). Once the first service is carried out on a patient, the progress bar is filled with a percentage that depends on the remaining operations to be performed ($x\%$). When the patient treatment is complete, the indicator becomes green and the progress bar is completely full (100%). The board on the right displays green and red indicators, respectively, showing the availability or non-availability of medical staff. This is the role of the RA.

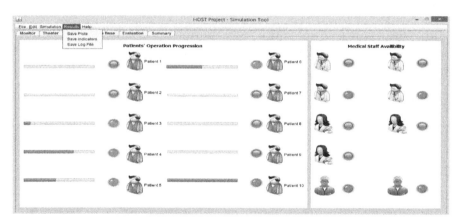

Figure 5.20. *Criteria evolution interface*

5.8.4.4. *Patient monitoring interface*

Thanks to the interaction between the SA, the TA and the RA, we get a visualization interface of the change in PES patients, which allows us to have an idea of the genuine ground situation, in real time (Figure 5.21). The progress bars give the SA a precise idea of the decisions to be made concerning staff allocation and the next patients to be treated. These progress bars clearly show the remaining care operations to be performed on the patients. In this way, the level of overcrowding is measured, not as a function of the amount of remaining patients at the waiting room, but as a function of the exact percentage of remaining tasks for patients who have already been taken care of. This interface provides the user with a general idea of the situation at PES just by visualizing the different percentages of remaining tasks to be performed.

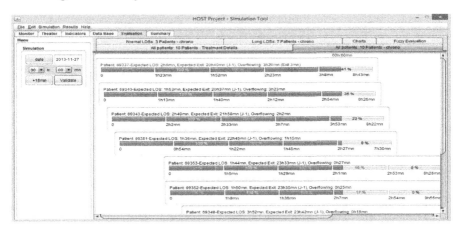

Figure 5.21. *Patient monitoring interface*

These bars evolve as a function of time. Their shallow parts represent care operation, which has already been performed and the empty ones show the percentage of remaining operations.

This model reveals the general remaining time. Excessive waiting delays reflect an overcrowding state and can be integrated in particular metrics, which can measure resource availability and service congestion.

5.9. Simulation and scheduling results: the SA's behavior

Using a real PES database, we have recovered a certain amount of information for each patient: age, sex, arrival mode, diagnosis, pathology and entry dates, which have allowed us to precisely inform the SA about the tasks to be scheduled and the medical staff to allocate.

The table below represents the patients' database. We have chosen a 10-patient scenario, which arrived at PES on November 27, 2013. For each patient, there is a corresponding pathology, an emergency degree and an arrival date. The PIA from SysCAOO platform sent the necessary information to the SA, in order to organize the schedule as a function of the necessary competencies to treat each pathology and their emergency degree. This information is detailed in the following Table 5.7:

id	patientID	name	pathology	level	arrivaldate	exitdate
8	8732	Patient7	concussion	3	2013-11-27 20:36:00	0000-00-00 00:00:00
10	1112	Patient10	concussion	2	2013-11-27 20:35:00	0000-00-00 00:00:00
7	5543	Patient2	gastroenteritis	1	2013-11-27 20:33:00	0000-00-00 00:00:00
9	9876	Patient8	gastroenteritis	1	2013-11-27 20:32:00	0000-00-00 00:00:00
6	9986	Patient1	compoundfracture	1	2013-11-27 20:31:00	0000-00-00 00:00:00
4	1223	Patient6	compoundfracture	5	2013-11-27 20:30:00	0000-00-00 00:00:00
5	3332	Patient3	gastroenteritis	3	2013-11-27 20:23:00	0000-00-00 00:00:00
1	4521	Patient9	compoundfracture	5	2013-11-27 20:22:00	0000-00-00 00:00:00
3	9875	Patient5	concussion	2	2013-11-27 20:16:00	0000-00-00 00:00:00
2	5646	Patient4	gastroenteritis	2	2013-11-27 20:10:00	0000-00-00 00:00:00

Table 5.7. *Table patients/pathologies*

For pathologies, we have chosen three types: concussion, gastroenteritis and compound fracture. A list of care tasks corresponds to each pathology, predefined by a care protocol. For the same pathology, corresponding to the same patient, there are many operations to be performed in parallel, or respecting precedence constraints. In this way, we attribute an execution order to each operation. For example, in Table 5.8, gastroenteritis pathology is composed of six operations of order 1 (which must be executed in parallel), followed by a number 2 operation and finally a number 3 operation.

id	designation	glassgow	hydration	tempreture	heartrate	bloodpressure	respiratoryfrequency	cerebralscanner	bacteriologicaltest	radiothorax	neurologicaltest	fracturescore	bloodtest	skeletonradio
1	concussion	1	0	1	1	1	1	3	0	3	2	0	2	0
2	gastroenteritis	0	1	1	1	1	1	0	2	3	0	0	1	0
3	compoundfracture	0	0	1	1	1	1	0	0	0	0	1	2	1

Table 5.8. *Operation database, classified by pathology*

Table 5.9 provides details about the care operations that must be executed for treating patients, in order 1–13. The third column represents theoretical execution time for each operation (in seconds). For each line, there are corresponding competence percentages ($0 \le \theta_{i,j,k} \le 100\%$) necessary for performing care operations.

Id	Operation_name	Execution time (sections)	Doctor score	Nurse Score	Aux score	Radiologist
1	Glasscow	500	100%	30%	0	0
2	Hydration	1,000	100%	60%	0	0
3	Temperature	500	100%	100%	60%	30%
4	Heartrate	1,000	100%	100%	0	0
5	Bloodpressure	300	100%	100%	0	0
6	Respiratoryfrequency	600	100%	100%	0	0
7	Cererbralscanner	1,200	60%	30%	100%	0
8	Bacteriologicaltest	500	60%	30%	100%	0
9	Radiothorax	1,200	30%	30%	0	100%
10	Neurogicaltest	2,000	100%	30%	0	0
11	Fracturescore	500	100%	100%	0	0
12	Bloodtest	300	30%	30%	100%	0
13	Skeketonradio	1,000	30%	0	0	100%

Table 5.9. *Competencies table*

There are 13 competencies in this table (C_1 = "Glasgow", C_2 = "Hydrations", ..., C_{13} = "Skeletonradio") and four types of medical staff: doctor, nurse, auxiliary and radiologist, with different competence percentages $\theta_{i,j,k}$, respectively, corresponding to columns 4, 5, 6 and 7.

If $\theta_{i,j,k} = 100\%$, that medical staff member is chosen by the SA to execute that operation. However, if the chosen staff member is not available, then the SA chooses another member, with a lower competence level ($\theta_{i,j,k} = 60\%$ or $\theta_{i,j,k} = 30\%$). If $\theta_{i,j,k} = 0$, that care operation cannot be performed. For each operation, the care protocol demands a minimum competence level, and this threshold depends on the nature of the care operation.

The medical team of PES consists of two pediatricians, one intern, two pediatrics interns, one surgery intern, two to three nurses and two auxiliaries. This team handles 13 competencies $C_{i,j,k}$ with different levels of experience and each care operation to be performed demands one or more competencies.

Table 5.10 presents all patients, their corresponding care operations and the necessary competencies for executing them. For each patient, we have a maximum of care operations to be executed by medical staff members, and for their execution, a minimum level of competence is required. In the case at hand, we have six different competencies.

	Operation 1	Operation 2	Operation 3	Operation 4	Operation 5
Patient 1	C1x2	C1 and C2	C1 and C3	C1 and C2x2	C4, C5x2 and C6
Patient 2	C2 and C3	C2 and C3	C2	No operation	No operation
Patient 3	C3x2	C3	No operation	No operation	No operation
Patient 4	C4x2	C5 and C6	C6x2	C4x2	C1 and C2
Patient 5	C2x2	C5	C5 and C6	C4 and C5	C3
Patient 6	C1	C4	C6	No operation	No operation
Patient 7	C6x2	C1	C5 and C6	C3	No operation
Patient 8	C3 and C5	C2 and C5	C3 and C6	C6	No operation
Patient 9	C5	C4	C1	No operation	No operation
Patient 10	C4	C4 and C5	C1 and C2	C4	No operation

Table 5.10. *Competence table*

The information reported in Table 5.10 is transmitted to the SA. This applies the EA for scheduling and allocating medical teams to multiple competencies.

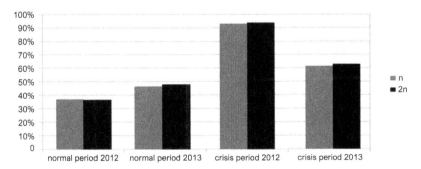

Skill 1															

Figure 5.22. *Gantt chart. For a color version of the figure, see www.iste.co.uk/zgaya/logistics.zip*

The Gantt chart shown in Figure 5.22 represents the scheduling of multiple competence tasks applied to 10 patients who await treatment. The schedule considered acknowledges classical constraints that we find in the specialized literature (precedence constraints and resource constraints), in addition to new constraints defined previously (section 5.2.5).

On the whole, the algorithm is capable of providing better satisfactory solutions than all the possible solutions that could be produced by PES practitioners. However, the convergence of the algorithm we use depends on many parameters. Our purpose is to analyze the impact of some of the parameters of this algorithm, in order to define the best configuration to adopt. In each case, the best solution in the population is chosen as representative for each test. The maximum number of generations in our EA is a key parameter for the method.

Figure 5.23. *The impact of generation numbers on four instances*

Figure 5.23 introduces the results of the variation in the number of generations between n and 2n. The ordinate axis represents the aggregative objective function to be minimized, the calculated sum of the four established criteria. The columns in gray represent the results when the number of generations is fixed as *n* and those in black represent the results when the number of generations is fixed as *2n*. The figure shows that the execution of the algorithm with 2n generations does not necessarily find better solutions. Indeed, after analyzing the database provided by PES, we have identified four instances corresponding to a crisis period in 2012, a crisis period in 2013, a normal period in 2012 and a normal period in 2013. Over these four instances, 2n generations have lowered the fitness function on three occasions, in comparison with n generations. This may be explained by the complexity of the problem for the cases considered and the existence of divisible and non-divisible care operations.

On the other hand, the increase in the number of maximum iterations has an influence. The execution time may increase by at least 10% if the maximum number of iterations is doubled. This could be an inconvenience for certain ES because they seek to obtain a task schedule as soon as possible.

The probability of applying local research (allocation process) can be a key factor in our approach. Table 5.11 presents the effects on the four instances with different p_m values. These values are considered for testing if a strong probability could improve the algorithm. Columns 0.1, 0.4, 0.5 and 0.6 present the tested values of p_m. Each column contains two sub-columns: execution time (CPU) in seconds and the quality of the solution (Cost). The default value for p_m is 0.1. Results show that an increase in the probability of calling local research is not always beneficial. This can be explained by the fact that the frequent call for local research does not necessarily improve the best solution in the population.

	0.1		0.4		0.5		0.6	
	CPU	Cost	CPU	Cost	CPU	Cost	CPU	Cost
2012 Normal Situation	55.318	0.3287	53.903	0.3639	55.749	0.3319	51.099	0.3618
2013 Normal Situation	74.431	0.6073	77.871	0.6267	61.722	0.6226	83.289	0.6207
2012 Critical Situation	93.689	0.9067	102.6	0.9318	103.13	0.9500	98.456	0.9458
2013 Critical Situation	151.66	0.4525	143.61	0.4935	200.5	0.4648	143.32	0.4664

Table 5.11. *The impact of p_m on the four instances*

Figure 5.24 shows a comparison between patient treatment time with and without using the SysCAOO system. This time period corresponds to the time spent by each patient at PES. Real treatment time is deduced from the database which has been provided to us. The graphic shows a time decrease in the amount of time spent by patients at PES. The last green vertical bar on the figure represents the average duration of treatment for 10 patients. The comparison shows a reduction in the average time spent at PES and an improvement in criterion Cr_3 .

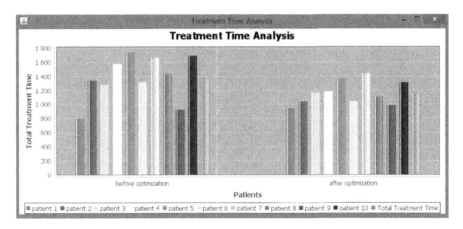

Figure 5.24. *Analysis of the amount of time spent by each patient at PES. For a color version of the figure, see www.iste.co.uk/zgaya/logistics.zip*

The second criterion that we have chosen for evaluating SysCAOO is the global waiting time of patients: Cr_2 .

The graph in Figure 5.25 shows a comparison between the waiting time of patients using our approach and the waiting time we deduced from real databases provided by PES. The graph reveals that, using our approach, the average waiting time is 40 minutes, compared to 3 hours in crisis periods.

5.9.1. *The impact of dynamic orchestration on Workflow*

In order to evaluate the impact of dynamic orchestration on Workflow, we have chosen the real scenario of a tense day, which took place at PES on February 7, 2013. Throughout this day, PES received around 119 patients, registering a remarkable annual activity peak. The difficulty lies in smoothing these activity peaks and reducing patients' waiting time.

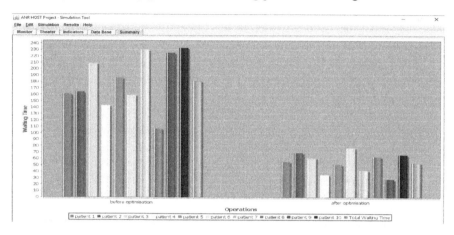

Figure 5.25. *Waiting time analysis. For a color version of the figure, see www.iste.co.uk/zgaya/logistics.zip*

To start the simulation, we define a patient arrival flow and calculate the necessary resources for executing care operations, taking into account the competencies of medical staff and resource availability.

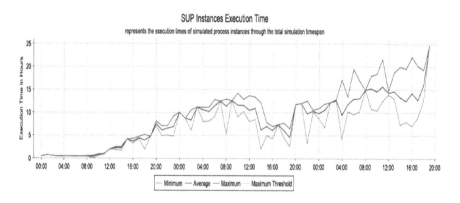

Figure 5.26. *Average waiting time with static orchestration. For a color version of the figure, see www.iste.co.uk/zgaya/logistics.zip*

The results obtained represent curves displaying overall minimum, maximum and average waiting time of patients at PES. In the first place, we introduce the results given by the SA, who is in charge of schedules using an evolutionary approach, with static orchestration. Then, we show that the communication of SA with other agents in order to dynamically orchestrate patient circuit Workflow makes it possible to reduce waiting time.

According to Figure 5.26, we observe that, on this overcrowding day, the average waiting time differs. The average waiting time is estimated at 10 hours. Medical staff are therefore incapable of facing the increase in patient flow arriving at PES.

Figure 5.27. *Average waiting time with dynamic orchestration.*
For a color version of the figure, see www.iste.co.uk/zgaya/logistics.zip

Figure 5.27 shows that the optimization approach based on the alliance between scheduling and MAS approaches has made it possible to smooth up activity peaks at PES and to minimize patient's average waiting time. Indeed, the SA plays the role of orchestra director, coordinating his actions with the other agents in order to ensure a dynamic orchestration adapted to the real situation of PES. In this case, the waiting time fluctuates between 1 and 4 hours. Nevertheless, we observe that there is always a midnight activity peak that disturbs PES.

The previous orchestration actions have not succeeded in absorbing all the PES activity peaks. Due to this, it is necessary to resort to negotiation between the agents of the formed coalition in order to dynamically re-orchestrate the Workflow of remaining patients. This negotiation is based on considering additional examination results (biological tests, X-ray, ecography, RMI, etc.) when making decisions about the patient's orientation.

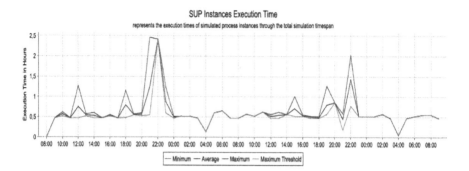

Figure 5.28. *Average waiting time with dynamic re-orchestration.*
For a color version of the figure, see www.iste.co.uk/zgaya/logistics.zip

The results obtained after dynamic re-orchestration are shown in Figure 5.28. The average waiting time is reduced and fluctuates between 30 minutes and 2 hours 30 minutes, with an activity peak between 6 pm and midnight.

5.10. Conclusion

In this chapter, we have introduced a mathematical formulation of criteria, constraints and decision variables, which acknowledge the real functioning of a PES. A multi-agent architecture has been proposed in order to ensure collaborative management between the different actors of a health center. In the behavior of these agents, we have integrated optimization algorithms capable of scheduling and orchestrating the patient circuit workflow. Different approaches have been grouped in a three-layer architecture, ensuring the likely course of the implemented system. In this chapter, we have presented simulation results using real PES databases, which prove the relevance of the approaches proposed in this chapter. We have simulated scenarios using the developed system SysCAOO, based on communicating agents. The purpose is to manage patient circuits and to improve PES care quality. The SysCAOO is provided with scheduling and orchestrating approaches which, in the case of overcrowding, make it possible to reduce pressure on medical staff and to shorten the waiting time. Using real 2011–2012–2013 databases, we have performed a fine analysis of different indicators which could generate a tense situation at PES, and a subsequent activity peak. Many simulation results have proved the efficiency of the alliance between optimization and MAS. The strength of our system relies on the formation of agent coalitions and their negotiation protocols, in order to make orchestration decisions and thus improve the waiting time during patient care, as well as to face the hazards that may come up at PES.

General Conclusion and Perspectives

Logistics engineering in the fields of health, transport or crisis management is a blooming sector that aims to efficiently employ technical and computerized means in order to optimize time management, to limit error risk and to anticipate disturbances, especially in these fields where the human factor is strongly present. The current state of these logistics systems can be described by precise indicators associated with internal factors, such as an increase in the tasks to be performed, a rise in case complexity, resource shortage, waiting time, etc. Besides the quick and regular augmentation of data on large-scale logistics networks and the imminent need to control and analyze this information, it demands the presence of management and optimization software at the core of the organizations involved. However, a multitude of varied tasks can concern data that may be diffuse, dynamic and vague. In this book, we have suggested using patient and medical staff information logistics flows for conceiving, optimizing and implementing a multiple competence care task orchestration collaborative system. This system has integrated a collaborative workflow (introduced in Chapter 4), containing diverse types of modules which serve for achieving different processes (complementary, rational and interactive), by exhaustively gathering pediatric emergency services (PES) patients' information.

An agent-based architecture represents a pertinent modeling approach for better flexibility, coherence, openness and interaction between autonomous and rational entities, in order to guarantee the robustness of logistics systems in the face of risks. In fact, we suggest providing agents with such an architecture by optimizing and scheduling roles and behavior, in order

Conclusion written by Hayfa ZGAYA and Slim HAMMADI.

to better manage the global functioning of logistics systems. In the field of health, thanks to agent interaction, we have sharpened the dynamic orchestration approach of workflow-modeled care processes, in order to improve the values of the system's ongoing optimization criteria. Therefore, we have proposed a three-layer generic architecture that can be customized. In this book, we have detailed the SysCAOO system, based on this three-layer architecture. This consists of a first mathematical and algorithmic layer, a second layer of communicating agents and a last layer representing experimentation ground, in which operational processes are modeled by a collaborative workflow approach.

SysCAOO was developed for improving PES patient care, through the optimization of human and material resources. It consists of a set of tailor-made agents, with different and complementary characteristics, customized to fit patient's needs. The number of agents in the system is not determined beforehand. Thanks to its flexibility, it is possible to add, suppress or modify agents, without altering the functioning of the overall system or the working of each component.

The "agentification" of SysCAOO, that is to say, its modeling in three autonomous rational and interactive layers, provides great coherence, evolution and complementarity. These aspects perfectly fit the health field, and PES in particular, because information is not sufficiently precise, readable or accessible for perfect resource management. Indeed, data are often homogeneous, diffuse, vague and redundant.

Research works and advances in the field of health introduced in this book are integrated in the context of the ANR HOST (2012–2015) project: "Hospital: Optimization, Simulation and Tension Avoidance", particularly in its Tecsan 2011 program: Health Technology and Autonomy. We have relied on the experience of emergency doctors (HOST project partners) as well as data collected between 2011 and 2013, during our visits to PES at Lille's Main Hospital Center (CHRU de Lille).

In Chapter 5, we simulated and tested many scenarios, with the purpose of evaluating SysCAOO. In order to dynamically schedule and orchestrate workflow models, we encouraged interaction between system agents, and consequently improved ongoing PES patient care. Finally, we compared the contributions of dynamic orchestration and static orchestration for the improvement of optimization criteria, such as waiting time.

This book introduces many scientific and technical perspectives:

1) the improvement of global system modeling, under the form of autonomous, rational and interactive entities, by integrating dynamic learning algorithms;

2) the enrichment of the implementation of each system agent, knowing that an agent can – by itself – represent a society of agents (holonic agent), as well as the definition and modeling of its internal architecture;

3) the reinforcement of protocol implementation (collaboration and negotiation) between different operational processes in the logistics system (agents and workflow);

4) solving a particular health problem may demand forming coalitions or agent sub-sets (as well as clearly specifying this particular process of coalition formation);

5) the integration of the paradigm of connected objects and big data within logistics systems. These objects, used by different agents for managing and optimizing the system, may be connected to many resources in different logistics fields, especially in the healthcare domain, such as blood samples, hospital beds and defibrillators.

Bibliography

[ABO 11] ABO-HAMAD W., ARISHA A., "Simulation-optimisation methods in supply chain applications: a review", *Irish Journal of Management*, vol. 30, pp. 95–124, 2011.

[ALV 05] ALVIN P., MARCELLI D., *Médecine de l'adolescent*, Masson, January 2005.

[ARM 02] ARMENGAUD D., "Le quiproquo des urgences pédiatriques", *Enfances & Psy*, vol. 2, pp. 10–16, 2002.

[ASP 03] ASPLIN B., MAGID D., RHODES K. *et al.*, "A conceptual model of emergency department crowding", *Annals of Emergency Medicine*, vol. 42, pp. 173–180, 2003.

[AUB 03] AUBERT N., *Le Culte de l'urgence: La société malade du temps*, Flammarion, 2003.

[BEA 99] BEAMON B.M., "Measuring supply chain performance", *International Journal of Operations and Production Management*, vol. 19, pp. 275–292, 1999.

[BEL 03] BELLOU A., DE KORWIN J.-D., BOUGET J. *et al.*, "Place des services d'urgences dans la régulation des hospitalisations publiques", *La Revue de Médecine Interne*, vol. 24, pp. 602–612, 2003.

[BEN 12] BENOIT C., *Manager un établissement de santé: la logistique au service de l'humain*, Gereso, 2012.

[BEN 14a] BEN CHEIKH S., HAMMADI S., TAHON C., "Based-agent distributed architecture to manage the dynamic multi-hop ridesharing system", *IEEE 13th International Symposium on Network Computing and Applications (NCA 2014)*, Cambridge, MA, USA, pp. 101–104, 2014.

[BEN 14b] BEN OTHMAN S., ZOGHLAMI N., HAMMADI S., "Adaptive collaborative agent-based system for crisis management", *2014 IEEE/WIC/ACM International Joint Conferences on Intelligent Agent Technology*, Varsovie, Pologne, pp. 151–158, 2014.

[BEN 14c] BEN OTHMAN S., ZOGHLAMI N., HAMMADI S. *et al.*, "Dynamic patients scheduling in the pediatric emergency department", *18th International Conference on Circuits, Systems, Communications and Computers*, Santorini Island, Greece, pp. 482–487, 2014.

[BEN 15a] BEN OTHMAN S., HAMMADI S., QUILLIOT A., "Agent-based evolutionary algorithm for multi-skill health care tasks scheduling", *Proceedings of the XI Metaheuristics International Conference*, Agadir, Morocco, pp. 105–111, 2015.

[BEN 15b] BEN OTHMAN S., HAMMADI S., QUILLIOT A., "Agents endowed with uncertainty management behaviors to solve a multi-skill health care tasks scheduling", *Proceedings of the 17th International Symposium on Health Information Management Research*, York, England, pp. 145–151, 2015.

[BEN 15c] BEN OTHMAN S., HAMMADI S., QUILLIOT A. *et al.*, "Health care decision support system for the pediatric emergency department management", *Proceedings of the 15th World Congress on Health and Biomedical Informatics*, São Paulo, Brazil, pp. 305–309, 2015.

[BEN 15d] BEN OTHMAN S., ZOGHLAMI N., HAMMADI S. *et al.*, "Multi-objective evolutionary for multi-skill health care tasks scheduling", *Proceedings of the 15th IFAC/IEEE/IFIP/IFORS Symposium/Information Control Problems in Manufacturing*, Ottawa, Canada, pp. 704–709, 2015.

[BOI 14] BOISGUÉRIN B., VALDELIÈVRE H., "Urgences: la moitié des patients restent moins de deux heures, hormis ceux maintenus en observation", Études et résultats, DREES, no. 889, July 2014.

[BOR 05] BORDINI R., DASTANI M., DIX J. *et al.*, *Multi-agent Programming Languages, Platforms and Applications*, International Book Series, Springer, 2005.

[BOT 05] BOTTA-GENOULAZ V., Principe et méthodes pour l'intégration et l'optimisation du pilotage des systèmes de production et des chaînes logistiques, thesis, INSA, University of Lyon, 2005.

[BOU 14] BOUSSELMI A., ZGAYA H., BOURDEAUD'HUY T. *et al.*, "A based dynamic role-enabled multi-agent information system", *COOS 2014: the 2nd AAMAS Workshop on Collaborative Online Organizations*, pp. 34–41, 2014.

[BRA 87] BRATMAN M., *Intention, Plans, and Pratical Reason*, Harvard University Press, Cambridge, MA, 1987.

[BRE 13] BRÉANT K., Analyse du recours au service des urgences pédiatriques du Havre par les médecins généralistes, PhD thesis, Faculty of Medicine and Pharmacy, Rouen University, p. 75, 2013.

[BRU 14] BRUNO G., GENOVESE A., PICCOLO C., "The capacitated Lot Sizing model: a powerful tool for logistics decision making", *International Journal of Production Economics*, vol. 155, pp. 380–390, 2014.

[CAB 14] CABEY W.V., MACNEILL E., WHITE L.N., "Frequent pediatric emergency department use in infancy and early childhood", *Pediatric Emergency Care*, vol. 30, no. 10, pp. 710–717, 2014.

[CAR 14] CARTER E., POUCH S., LARSON E., "The relationship between emergency department crowding and patient outcomes: a systematic review", *Journal of Nursing Scholarship*, vol. 46, pp. 106–115, 2014.

[CER 13] CEREZO N., MONTAGNAT J., BLAY-FORNARINO M., "Computer-assisted scientific workflow design", *Journal of Grid Computing*, vol. 11, pp. 585–612, 2013.

[CHE 04] CHEVASSUS M., "Enterprise application integration: EAI", *Techniques de l'ingénieur*, pp. 1–20, April 2004.

[CHI 12] CHINOSI M., TROMBETTA A., "BPMN: An introduction to the standard", *Computer Standards and Interfaces*, vol. 34, pp. 124–134, 2012.

[CHR 11] CHRISTOPHER M., *Logistics and Supply Chain Management*, Financial Times/Prentice Hall, 2011.

[CIR 13] CIRULIS A., GINTERS E., "Augmented reality in logistics", *Procedia Computer Science*, vol. 26, pp. 14–20, 2013.

[COH 01] COHEN N., "Au delà du supply chain management: le Global Fulfillment", *Logistique & Management*, vol. 9, pp. 89–92, 2001.

[COL 02] COLLETTE Y., SIARRY P., *Optimisation Multiobjectif*, Eyrolles, Paris, 2002.

[COL 07] COLOMBIER G., La prise en charge des urgences, Rapport d'information – Assemblée Nationale, p. 575, 2007.

[COO 04] COOKE M., FISHER J., DALE J. *et al.*, Reducing attendances and waits in emergency departments: a systematic review of present innovations, Report National Co-ordinating Centre for NHS Service Delivery and Organisation R&D (NCCSDO), London, 2004.

[COU 03a] COUANAU R., L'organisation interne de l'hôpital, Rapport de la Commission Des Affaires Culturelles, Familiales et Sociales de l'Assemblée Nationale, 2003.

[COU 03b] COURTOIS A., PILLET M., MARTIN-BONNEFOUS C., *Gestion de production*, Editions d'Organisation, 2003.

[CRO 00] CROOM S., ROMANO P., GIANNAKIS M., "Supply chain management: an analytical framework for critical literature review", *European Journal of Purchasing & Supply Management*, vol. 6, pp. 67–83, 2000.

[DAK 11] DAKNOU A., Architecture distribuée à base d'agents pour optimiser la prise en charge des patients dans les services d'urgences en milieu hospitalier, PhD thesis, LAGIS, Ecole centrale de Lille, 2011.

[DER 00] DERLET R., RICHARDS J., "Overcrowding in the nation's emergency departments: complex causes and disturbing effects", *Annals of Emergency Medicine*, vol. 35, pp. 63–68, 2000.

[DEV 97] DEVICTOR D., COSQUER M., SAINT-MARTIN J., "L'accueil des enfants aux urgences: résultats de deux enquêtes nationales", *Archives de Pédiatrie*, vol. 4, pp. 21–26, 1997.

[DOE 07] DOERNER K.F., REIMANN M., "Logistics of health care management, part special issue: logistics of health care management", *Computers & Operations Research*, vol. 34, pp. 621–918, 2007.

[DON 06] DONIEC A., Prise en compte des comportements anticipatifs dans la coordination multi-agent: application à la simulation de trafic en Carrefour, University of Valenciennes and Hainaut-Cambrésis, Valenciennes, 2006.

[DOU 98] DOUMEINGTS G., VALLESPIR B., CHEN D., "GRAI grid decisional modelling", in BERNUS P., MERTINS K., SCHMIDT G. (eds), *Handbook on Architectures of Information Systems*, Springer, pp. 313–337, 1998.

[DRÉ 03] DRÉO J., PÉTROWSKI A., SIARRY P. *et al.*, *Métaheuristiques pour l'optimisation difficile*, Eyrolles 2003.

[ELK 12] ELKHALDI-MKAOUAR I., Application de la Différentiation Automatique pour l'identification, l'optimisation et l'étude de sensibilité dans quelques probèmes mécaniques, University of Lorraine, 2012.

[ELL 15] ELLBRANT J., ÅKESON J., ÅKESON P.K., "Pediatric emergency department management benefits from appropriate early redirection of nonurgent visits", *Pediatric Emergency Care*, vol. 31, pp. 95–100, 2015.

[FAB 04] FABBE-COSTES, N., ROMEYER, C., "The traceability of care activities by SIH: inventory", *Logistics and Management*, Special Issue 2004 "Hospital Logistics", pp. 119–133, 2004.

[FAR 11] FAROUK I.E.I., ABDENNEBI T., FOUAD J., "Modeling and simulation of hospital supply chain: state of the art and research perspectives", *Presented at the International Conference on Logistics (LOGISTIQUA)*, pp. 287–291, 2011.

[FEE 07] FEE C., WEBER E., "Identification of 90% of patients ultimately diagnosed with community-acquired pneumonia within four hours of emergency department arrival may not be feasible", *Annals of Emergency Medicine*, vol. 5, pp. 553–559, 2007.

[GAL 09] GALVIS-NARINOS F., "Le système de santé des États-Unis", *Pratiques et organisation des Soins*, vol. 40, pp. 309–315, 2009.

[GAN 99] GANESHAN R., JACK E., MAGAZINE M.J. *et al.*, "A taxonomic review of supply chain management research", in TAYUR S., GANESHAN R., MAGAZINE M. (eds), *Quantitative Models for Supply Chain Management*, International Series in Operations Research & Management Science, Springer US, pp. 840–879, 1999.

[GEN 05] GENIN P., LAMOURI S., THOMAS A., "Planification avancée: APS", *Techniques de l'ingénieur*, pp. 1–12, April 2005.

[GEN 10] GENIN T., AKNINE S., "Coalition formation strategies for self-interested agents in task oriented domains", *IEEE/WIC/ACM International Conference on Web Intelligence and Intelligent Agent Technology*, pp. 205–212, 2010.

[GEN 13] GENTIL S., "Les "agencements organisationnels" des situations perturbées: la coordination d'un bloc opératoire à la pointe de la rationalisation industrielle", *Communiquer*, vol. 8, pp. 65–80, 2013.

[GER 12] GEROIMENKO V., "Augmented reality technology and art: The analysis and visualization of evolving conceptual models", *Presented at the 2012 16th International Conference on Information Visualisation (IV)*, IEEE4, pp. 445–453, 2012.

[GOU 91] GOURGAND M., KELLER P., "Conception d'un environnement de modélisation des systèmes de production", *Congrès International de Génie Industriel*, pp. 191–203, 1991.

[GRE 05] GREENWOOD D., JADE Web Service Integration Gateway (WSIG), Whitestein Technologies, Jade Tutorial, AAMAS 2005.

[GUD 10] GUDEHUS T., KOTZAB H., *Logistik: Grundlagen – Strategien – Anwendungen, (Comprehensive Logistics)*, Springer, Berlin-Heidelberg, 2010.

[HAL 91] HALES K., MANDY L., OVUM LTD., *Workflow Management Software: The Business Opportunity*, Ovum Ltd, 1991.

[HAM 12a] HAMMADI S., ZGAYA H., "Système d'information à base d'agents pour la recherche et la composition des services pour l'aide à la mobilité", in HAMMADI S., KSOURI M. (eds), *Ingénierie Du Transport et Des Services de Mobilité Avancés*, Hermes Science-Lavoisier, Paris, pp. 51–106, 2012.

[HAM 12b] HAMMADI S., ZGAYA H., "An agent based information system for searchning and creating mobility-aiding services", in HAMMADI S., KSOURI M. (eds), *Advanced Mobility and Transport Engineering*, ISTE, London and John Wiley & Sons, New York, pp. 31–79, 2012.

[HIN 11] HINCAPIE M., "An introduction to augmented reality with applications in aeronautical maintenance", *Presented at the 2011 13th International Conference on Transparent Optical Networks (ICTON)*, IEEE, pp. 1–4, 2011.

[HOH 06] HOHMANN C., "La convergence qualité/logistique", *Management*, pp. 5–6, July 2006.

[HOO 08] HOOT N.R., ARONSKY D., "Systematic review of emergency department crowding: causes, effects and solutions", *Annals of Emergency Medicine*, vol. 2, pp. 126–36, 2008.

[HSI 06] HSIEH F.-S., "Analysis of contract et in multi-agent systems", *Automatica*, vol. 42, pp. 733–740, 2006.

[HUA 05] HUANG S.H., SHEORAN S.K., KESKAR H., "Computer-assisted supply chain configuration based on supply chain operations reference (SCOR) model", *Computers & Industrial Engineering*, vol. 48, pp. 377–394, 2005.

[HUE 11] HUE V., DUBOS F., PRUVOST I. *et al.*, "Organisation et moyens des urgences pédiatriques: enquête nationale française en 2008", *Archives Pédiatriques*, vol. 18, pp. 42–48, 2011.

[IOM 01] INSTITUTE OF MEDICINE, *Crossing the Quality Chasm: A New Health System for the 21st Century*, National Academies Press, 2001.

[JAC 12] JACQUES T., *Recherche Opérationnelle – Tome 1: Programmation linéaire. Optimisation combinatoire. Programmation dynamique. Graphes. Métaheuristiques*, Ellipses, 2012.

[JER 12] JERIBI K., Conception et réalisation d'un système de gestion de véhicules partagés: de la multimodalité vers la co-modalité, PhD thesis, Ecole Centrale de Lille, 2012.

[JOU 05] JOUENNE T., D'une Logistique Fragmentée à la Logistique Systémique, Centrale Paris, 2005.

[KAD 12] KADDOUCI A., Optimisation des fux logistiques: vers une gestion avancée de la situation de crise, LAGIS UMR CNRS 8 219, Ecole Centrale de Lille, 2012.

[KAD 13] KADRI F., PACH C., CHAABANE S. *et al.*, "Modelling and management of strain situations in hospital systems using an ORCA approach", *5th IESM Conference*, Rabat, Morocco, pp. 1–9, 2013.

[KAD 14] KADRI F., HARROU F., CHAABANE S. *et al.*, "Time series modelling and forecasting of emergency department overcrowding", *Journal of Medical Systems*, vol. 38, pp. 1–20, 2014.

[KAL 04] KALINICHENKO A.V., JAVORONKOV E.P., CHRISTENKO E.L. *et al.*, "Logistic management in public health", *8th Russian-Korean International Symposium on Science and Technology,* pp. 234–236, 2004.

[KAM 98] KAMATH M.U., Improving correctness and failure handling in workflow management systems, PhD thesis, University of Massachusetts, 1998.

[KAM 02] KAMINSKY P., SIMCHI-LEVI E., *Designing and Managing the Supply Chain: Concepts, Strategies, and Case Studies*, McGraw-Hill/Irwin, October 2002.

[KAM 07] KAMOUN M.A., Conception d'un système d'information pour l'aide au déplacement multimodal: Une approche multi-agents pour la recherche et la composition des itinéraires en ligne, Ecole Centrale de Lille, AGIS, 2007.

[KEH 11] KEHE W., HUAN Z., GANG M., "Design and implementation of ARIS methodology-based process modeling and management platform", *Energy Procedia*, pp. 430–436, 2011.

[KOU 06] KOUVELIS P., CHAMBERS C., WANG H., "Supply chain management research and production and operations management: review, trends, and opportunities", *Production and Operations Management*, vol. 15, pp. 449–469, 2006.

[LAC 13] LACROIX B., MATHIEU P., KEMENY A., "Formalizing the construction of populations in multi-agent simulations", *Engineering Applications of Artificial Intelligence*, vol. 26, pp. 211–226, 2013.

[LAU 04] LAURAS M., Méthodes de diagnostic et d'évaluation de performance pour la gestion de chaînes logistiques: application à la coopération maison-mère – filiales internationales dans un groupe pharmaceutique et cosmétique, PhD thesis, National Polytechnic Institute of Toulouse, 2004.

[LEE 13] LEE S., LEE B., LEE A. *et al.*, "Augmented reality intravenous injection simulator based 3D medical imaging for veterinary medicin", *Veterinary Journal*, vol. 196, pp. 197–202, 2013.

[LEG 09] LEGRAND J.M., Report for hospital reform relative to patients, health and territories, report no. 1441, National Assembly, February 2009.

[LEM 08] LEMOINE D., Modèles génériques et méthodes de résolution pour la planification tactique mono-site et multi-site, Blaise Pascal University, 2008.

[LIÈ 07] LIÈVRE P., *La logistique, La Découverte*, Paris, 2007.

[LIU 11] LIU H., LEMBARET Y., CLIN D. *et al.*, "Comparison between collaborative business process tools", *2011 Fifth International Conference on Research Challenges in Information Science (RCIS)*, pp. 1–6, 2011.

[LOU 01] LOURENÇO H.R.D., "Supply chain management: an opportunity for metaheuristics", Economics Working Papers, Department of Economics and Business, Pompeu Fabra University Barcelone, 2001.

[LUM 98] LUMMUS R., VOKURKA R., ALBERT K., "Strategic supply chain planning", *Production and Inventory Management Journal*, vol. 39, pp. 49–58, 1998.

[MAC 02] MACINTYRE B., LOHSE M., BOLTER J.D. *et al.*, "Integrating 2-D video actors into 3-D augmented-reality systems", *Presence*, vol. 11, pp. 189–202, 2002.

[MAR 04] MARQUARDT W., NAGL M., "Workflow and information centered support of design processes – the IMPROVE perspective", *Computers & Chemical Engineering*, vol. 29, pp. 65–82, 2004.

[MAR 05] MAROUSEAU G., "Le système logistique, facteur-clé du succés des cybermarchés", *Logistique & Management*, vol. 13, pp. 9–20, 2005.

[MAR 08] MARCON E., GUINET A., TAHON C., *Gestion et Performance des Systèmes Hospitaliers*, Hermes Science-Lavoisier Paris, 2008.

[MAT 10] MATHE H., TIXIER D., *La Logistique*, Presses Universitaires de France, 2010.

[MAZ 10] MAZIER A., Optimisation Stochastique pour la gestion des lits d'hospitalisation sous incertitudes, PhD thesis, Ecole Nationale Supérieure des Mines de Saint-Etienne, 2010.

[MCK 02] MCKEE M., HEALY J., *Hospitals in a Changing Europe*, Open University Press, 2002.

[MEL 14] MELI CHRISTOPHER L., KHALIL I., TARI Z., "Load-sensitive dynamic workflow re-orchestration and optimisation for faster patient healthcare", *Computer Methods and Programs in Biomedicine*, vol. 113, pp. 1–14, 2014.

[MEN 11] MENTZER J.T., DEWITT W., KEEBLER J.S. *et al.*, "Defining supply chain management", *Journal of Business Logistics*, vol. 22, pp. 1–25, 2011.

[MIN 98] MINVIELLE E., "Gérer la singularité à grande échelle: Le cas des patients hospitalisés", *Guest Hospital*, no. 373, pp. 129–145, 1998.

[MIN 08] MINISTÈRE DES AFFAIRES SOCIALES ET DE LA SANTÉ, Rapport de la commission de concertation sur les missions de l'hôpital, présidée par M. Gérard Larcher, report, April 2008.

[MOL 05] MOLINIÉ É., L'hôpital public en France, Bilan et perspectives (Etude du Conseil Economique et Social), 2005.

[MOR 11] MORRIS Z., BOYLE A., BENIUK K. *et al.*, "Emergency department crowding: towards an agenda for evidence-based intervention", *Emergency Medicine Journal*, vol. 6, pp. 460–466, 2011.

[MOS 09] MOSKOP J.C., SKLAR D., GEIDERMAN J., "Emergency department crowding, part 1-concept, causes, and moral consequences", *Annals of Emergency Medicine*, vol. 5, pp. 605–611, 2009.

[MÜH 02] MÜHLENET Z., MUEHLEN M., ZHAO J., "Tutorial workflow and process automation in the age of e-business", *Proceedings of Thirly-First Annual Hawaii International Conference on System Sciences*, USA, 2002.

[MÜL 06] MÜLLER I., KOWALCZYK R., BRAUN P., "Towards agent-based coalition formation for service composition", *IEEE/WIC/ACM International Conference on Intelligent Agent Technology*, Hong Kong, pp. 73–80, 2006.

[NAT 08] NATHAN L., TIMM M.D., MONA L. *et al.*, "Pediatric emergency department overcrowding and impact on patient flow outcomes", *Academic Emergency Medicine*, vol. 15, pp. 832–837, 2008.

[NEE 12] NEE A.Y.C., ONG S.K., CHRYSSOLOURIS G. *et al.*, "Augmented reality applications in design and manufacturing", *CIRP Annals – Manufacturing Technology*, vol. 61, pp. 657–679, 2012.

[NES 08] NESRINE Z., Optimisation à base d'agents communicants des flux logistiques pour la gestion de crises, Ecole Centrale de Lille, 2008.

[OSM 96] OSMAN I.H., LAPORTE G., "Metaheuristics: a bibliography", *Annals of Operations Research*, vol. 63, pp. 513–623, 1996.

[OTH 14] OTHMAN S.B., ZOGHLAMI N., HAMMADI S. *et al.*, "Adaptive collaborative agent-based system for crisis management", *2014 IEEE/WIC/ACM International Conference on Intelligent Agent Technology (IAT 2014)*, pp. 151–158, 2014.

[PIN 11] PINES J.M., HILTON J.A., WEBER E.J. *et al.*, "International perspectives on emergency department crowding", *Academic Emergency Medicine: Official Journal of the Society for Academic Emergency Medicine*, vol. 18, pp. 1358–1370, 2011.

[REI 05] REID P.P., DALE COMPTON W., GROSSMAN J.H, *Building a Better Delivery System: A New Engineering/Health Care Partnership*, National Academies Press, 2005.

[ROH 00] ROHDE J., MEYR H., WAGNER M., Die Supply Chain Planning Matrix, Publications of Darmstadt Technical University, Institute for Business Studies (BWL), 2000.

[ROT 01] ROTA-Franz K., THIERRY C., BEL G., "Gestion des flux dans les chaînes logistiques (supply chain management)", in BURLAT P., CAMPAGNE J.-P. (eds), *Performances Industrielles et Gestion Des Flux*, Hermes Science-Lavoisier Paris pp. 103–128, 2001.

[ROW 06] ROWE B.H., BOND K., OSPINA M., "Emergency department overcrowding in Canada: what are the issues and what can be done?", *Ottawa: Canadian Agency for Drugs and Technologies in Health*, vol. 9, pp. 641–645, 2006.

[SAA 10a] SAAD S., Conception et optimisation distribuée d'un système d'information des services d'aide à la mobilité urbaine basé sur une ontologie flexible dans le domaine de transport, Ecole Centrale de Lille, 2010.

[SAA 10b] SAAD S., ZGAYA H., HAMMADI S., "Knowledge management", PASI VIRTANEN, NINA HELANDER (eds), *Integral Part of Information and Communications Technology*, InTech, 2010.

[SAL 97] SALTMAN R.B., FIGUERAS J. (eds), European Health Care Reform: Analysis of Current Strategies, WHO Regional Publications, European Series, no. 72, p. 310, 1997.

[SAM 12] SAMU-URGENCES DE FRANCE, Comment garantir l'accès à des soins médicaux de qualité en urgence? Les premières Assises de l'Urgence (Rapport final), Paris, 2012.

[SCH 12] SCHUUR J.D., VENKATESH A.K., "The growing role of emergency departments in hospital admissions", *The New England Journal of Medicine*, vol. 367, pp. 391–393, 2012.

[SCH 13] SCHABACKER M., GERICKE K., SZÉLIG N. *et al.*, "Modelling and management of engineering", in SCHABACKER M., GERICKE K., SZÉLIG N. *et al.*, *Processes, Proceedings of the 3rd International Conference*, p. 203, 2013.

[SGH 11] SGHAIER M., Combinaison des techniques d'optimisation et de l'intelligence artificielle distribuée pour la mise en place d'un système de covoiturage dynamique, PhD thesis, Ecole Centrale de Lille, 2011.

[SHA 99] SHAPIRO J.F., "On the connections among activity-based costing, mathematical programming models for analyzing strategic decisions, and the resource-based view of the firm", *European Journal of Operational Research*, vol. 118, pp. 295–314, 1999.

[SHE 02] SHEER A.W., *ARIS – Business Process Modelling*, Springer, 2002.

[SIA 14] SIARRY P., *Métaheuristiques*, Eyrolles, 2014.

[SIM 99] SIMCHI-LEVI D., KAMINSKY P., SIMCHI-LEVI E., *Designing and Managing the Supply Chain: Concepts, Strategies and Case Studies*, Irwin/McGraw-Hill, 1999.

[SOC 01] SOCIÉTÉ FRANÇAISE DE MÉDECINE D'URGENCE, Critères d'évaluation des Services D'urgences, 2001.

[SOC 04] SOCIÉTÉ FRANÇAISE D'ANESTHÉSIE et DE RÉANIMATION, Recommandations pour l'organisation de la prise en charge des urgences vitales intra hospitalières, 2004.

[SOC 08a] SOCIÉTÉ FRANÇAISE DE MÉDECINE D'URGENCE, Compétences Infirmier(e) en médecine d'urgence (Référentiel), 2008.

[SOC 08b] SOCIÉTÉ FRANÇAISE DE MÉDECINE D'URGENCE, "Les français et l'accès aux soins urgents non programmés", *2ème Congrès de la SFMU*, Paris, France, 2008.

[SOH 13] SOHIER J., *La logistique*, Vuibert, 2013.

[STA 04] STAGNARA J., VERMONT J., DUQUESNE A. *et al.*, "Urgences pédiatriques et consultations non programmées: enquête auprès de l'ensemble du système de soins de l'agglomération lyonnaise", *Archives Pédiatriques*, vol. 11, pp. 108–114, 2004.

[STA 12] STANGER S.H.W., WILDING R., YATES N. *et al.*, "What drives perishable inventory management performance? Lessons learnt from the UK blood supply chain", *Supply Chain Management: An International Journal*, vol. 17, pp. 107–123, 2012.

[STE 09] STEAD W.W., LIN H.S., *Computational Technology for Effective Health Care: Immediate Steps and Strategic Directions*, National Academies Press, 2009.

[STO 03] STORMER H., Ein flexibles und sicheres agentenbasiertes Workflow-Management-System, PhD Thesis, University of Zurich, 2003.

[SUN 13] SUN B.C., HSIA R.Y., WEISS R.E. *et al.*, "Effect of emergency department crowding on outcomes of admitted patients", *Annals of Emergency Medicine*, vol. 61, pp. 605–611.e6, 2013.

[THO 96] THOMAS D.J., GRIFFIN P.M., "Coordinated supply chain management", *European Journal of Operational Research*, vol. 94, pp. 1–15, 1996.

[THO 00] THOMAS A., LAMOURI S., "Flux poussés: MRP et DRP", *Techniques de l'ingénieur*, AGL1, 1–12, 2000.

[TLA 09] TLAHIG H., Vers un outil d'aide à la décision pour le choix entre internalisation/externalisation ou mutualisation des activités logistiques au sein d'un l'établissement de santé: cas du service de sterilization, PhD thesis, Grenoble Institute of Technology, 2009.

[TOR 06] TORRES O.A.C., MORÁN F.V., *The bullwhip effect in supply chains: A review of methods, components and cases*, Palgrave Macmillan, 2006.

[TRZ 03] TRZECIAK S., RIVERS E.P., "Emergency department overcrowding in the United States: an emerging threat to patient safety and public health", *Emergency Medicine Journal*, vol. 20, pp. 402–405, 2003.

[UED 01] UEDA K., "Synthesis and emergence – research overview", *Artificial Intelligence in Engineering*, vol. 15, pp. 321–327, 2001.

[VAK 02] VAKHARIA A.J., "e-Business and supply chain management", *Decision Sciences*, vol. 33, pp. 495–504, 2002.

[VAL 08] VALLANCIEN G., Réflexions et propositions sur la gouverance hospitalière et le poste de président du directoire, report, July 2008.

[VAT 12] VATE-U-LAN P., "An augmented reality 3D pop-up book: The development of a multimedia project for English language teaching", *2012 IEEE International Conference on Multimedia and Expo (ICME)*, IEEE, pp. 890–895, 2012.

[VEN 95] VENKATRAMAN N., "Reconfigurations d'entreprises provoquées par les technologies de l'information", in SCOTT-MORTON M.S. (ed.), *L'entreprise Compétitive Au Futur*, Editions d'Organisation, 1995.

[VIN 04] VINCENT L., NEUBERT G., PELLEGRIN C. *et al.*, Synthèse des approches SCM existantes. Ecole Supérieure de Commerce de Saint-Etienne, COACTIS, Economie Appliquée de Grenoble – GAEL, 2004.

[WAN 15] WANG Z., Optimisation avancée pour la recherche et la composition des itinéraires comodaux au profit des clients de transport, Ecole Centrale de Lille, 2015.

[WEI 99] WEISS G., *Multiagent Systems*, The MIT Press, Cambridge, MA, 1999.

[WIT 12] WITMEUR R., DÉSIR D., HUT F., "La réforme Obama du système américain de soins de santé", *CRISP*, 2012.

[WOL 08] WOLF J., *The Nature of Supply Chain Management Research*, Springer, vol. 35, pp. 1–100, 2008.

[ZGA 07a] ZGAYA H., Conception et optimisation distribuée d'un système d'information d'aide à la mobilité urbaine: Une approche multi-agent pour la recherche et la composition des services liés au transport, Ecole Centrale de Lille, France, 2007.

[ZGA 07b] ZGAYA H., HAMMADI S., "Multi-agent information system using mobile agent negotiation based on a flexible transport ontology", *2007 International Conference on Autonomous Agents and Multiagent Systems (AAMAS'2007)*, pp. 1034–1036, 2007.

[ZGA 09] ZGAYA H., HAMMADI S., "Distributed optimization using the mobile agent paradigm through an adaptable ontology: Multi-operator services research and composition", in AHMED S., KARSITI M.N. (eds), *Multiagent Systems, Artificial Intelligence*, In-Tech, pp. 397–426, 2009.

[ZGA 10] ZHENG K., HAFTEL H.M., HIRSCHL R.B. *et al.*, "Quantifying the impact of health IT implementations on clinical workflow: a new methodological perspective", *Journal of the American Medical Informatics Association: JAMIA*, vol. 17, pp. 454–461, 2010.

[ZID 06] ZIDI K., Système Interactif d'Aide au Déplacement Multimodal (SIADM), Ecole Centrale de Lille et l'Université des Sciences et Technologies de Lille, 2006.

List of Authors

Inès AJMI
Ecole Centrale de Lille
France

Sarah BEN OTHMAN
Ecole Centrale de Lille
France

Slim HAMMADI
Ecole Centrale de Lille
France

Alain QUILLIOT
LIMOS, ISIMA
Aubière
France

Jean-Marie RENARD
CERIM/Lille 2
University of Health and Law
France

Hayfa ZGAYA
ILIS/Lille 2
University of Health and Law
France

Index

Printed in the United States
By Bookmasters